Building Measurement

Measurement of buildings is the core skill of the quantity surveyor. It underpins the procurement, management, delivery and subsequent commissioning of a completed building, and must now be completed using New Rules of Measurement 2 (NRM2).

In this much-needed new measurement textbook, measurement of the most common building elements is described using NRM2. Extensive worked examples including fully up-to-date hand-drawn diagrams and supporting take-off lists ensure that the reader develops a confidence in their ability to measure using NRM2 in practice.

A practical step-by-step approach is used to explain and interpret the detail of the specific Work Sections of NRM2, covering a broad range of different trades, including mechanical and electrical systems; external works; groundwork; masonry; joinery; and internal finishes.

Presuming no prior knowledge of measurement or NRM2, and fully up to date with current practice, including consideration of Building Information Modelling, this is the ideal text for students of measurement at HND or BSc level, as well as practitioners needing a crash course in how to apply NRM2.

Visit the eResource website: www.routledge.com/9781138838147

Andrew Packer is a Chartered Quantity Surveyor and Associate Head at the University of Portsmouth, UK. He has designed and delivered learning materials for professional practice, undergraduate and postgraduate courses for over thirty years. He is an external examiner for RICS accredited programmes in the UK and was responsible for introducing the first RICS accredited courses in the Russian Federation.

Building Measurement

New rules of measurement

Second edition

Andrew Packer

Routledge
Taylor & Francis Group

LONDON AND NEW YORK

Second edition published 2017
by Routledge
2 Park Square, Milton Park, Abingdon, Oxon OX14 4RN

and by Routledge
711 Third Avenue, New York, NY 10017

Routledge is an imprint of the Taylor & Francis Group, an informa business

First edition published by Addison Wesley Longman 1996

British Library Cataloguing-in-Publication Data
A catalogue record for this book is available from the British Library

Library of Congress Cataloging in Publication Data
Names: Packer, Andrew (Andrew D.), author.
Title: Building measurement : new rules of measurement / Andrew Packer.
Description: Second edition. | New York, NY : Routledge, 2016. | Includes
bibliographical references and index.
Identifiers: LCCN 2016007727| ISBN 9781138694026 (hardback : alk. paper) |
ISBN 9781138838147 (pbk. : alk. paper) | ISBN 9781315734613 (ebook : alk. paper)
Subjects: LCSH: Building – Estimates. | Buildings – Measurement. | Buildings –
Valuation. | Quantity surveying. | Building – Estimates – Standards – Great Britain.
Classification: LCC TH435 .P236 2016 | DDC 692/.5 – dc23
LC record available at http://lccn.loc.gov/2016007727

ISBN: 978-1-138-69402-6 (hbk)
ISBN: 978-1-138-83814-7 (pbk)
ISBN: 978-1-315-73461-3 (ebk)

Typeset in Baskerville
by Florence Production Ltd, Stoodleigh, Devon

Isobel Jane Kilgallon, 1951–2015

Contents

Front piece or preface

This book is intended for those new to the idea of building measurement and is written in a style that assumes no prior knowledge of the subject. Wherever necessary the text is supported by illustrations, examples and visual cues as an aid to interpretation. Even so, there are no shortcuts and the only way to get to grips with recording dimensions is to have a go. It is important to put pen to dimension paper and to be prepared to make mistakes. It is only by making mistakes that any of us are able to learn.

It is twenty years since the first edition of this text was published. During that time many things have changed, few more than the communication and IT technologies that we take for granted today. The very idea that anyone would attempt to record dimensions, prepare an Abstract and produce a Bill of Quantities using a traditional approach based on manual systems seems incongruous with the software and technology that is presently available. The Bill of Quantity production process of 1996 would have taken months; today the same is possible in minutes.

As appealing as the notion may seem, the idea that a Bill of Quantities will simply 'materialise' at the press of a button is misleading. Those new to the practice of preparing tender documentation will still need to be able to interpret construction contracts and understand the building procurement processes that underpin the financial management of construction projects. Equally important is the ability to be able to interpret drawn information and appreciate how buildings are constructed. Any attempt at an explanation that excludes a sound grasp of these basic principles would be of limited value.

The first four chapters of this text provide an introduction and give a framework for measurement and document preparation. All are based on a traditional approach to the production of a Bill of Quantity (BQ) and are written assuming the adoption of the New Rules of Measurement 2 (NRM2). In order to put this in context, Chapter 1 includes an explanation and review of the suite of documents that are available for use in conjunction with NRM2 (comprising NRM1 and NRM3). No introduction would be complete without mention of Building Information Modelling (BIM) and the preparation of a digital model that represents the completed building (Chapter 4). An online student measurement guide is also available to support Chapter 2.

Each of the following eleven chapters adopts a practical step-by-step approach in order to explain and interpret the Work Sections of NRM2. Frequently this information was easier to convey by the use of sketches, diagrams and tables. In order to further assist interpretation, an outline of the relevant construction technology is also included. The majority of the worked examples that follow each chapter are based on traditional low-rise UK residential property, with three exceptions: basements, concrete-framed buildings and steel-framed buildings. Chapters 5 to 15 each include at least one worked sample take-off based on drawings and specification in accordance with the most recent building regulation and contemporary environmental standards. The text of each chapter includes NRM2 annotations and take-off lists, together with diagrams illustrating presentation and approach.

Primarily intended for undergraduate students studying built environment, construction and quantity surveying courses, the text should be equally applicable to BTEC and Extended Diploma students. It is also hoped that it will prove a useful reference for all those involved in the management and administration of construction work.

Despite the rigours of NRM2, it is not unusual for custom and practice to vary between individual surveyors and from one office to another. With this in mind, the reader is advised that the techniques described here need not be taken as exclusive; wherever possible a best practice approach has been adopted. Where adjustments have been made in the following sample take-off (deductions), these would normally be recorded using a red pen. Since this text is not printed in colour these are only evident by inclusion of the abbreviation '*Ddt*'.

A.D. Packer, January 2016

Acknowledgements

There have been many people who have assisted in the preparation of this text; to all of them I am extremely grateful. Particular thanks are due to my colleague Dr Mark Danso-Amoako for his help, advice and guidance with the text, proofreading, worked examples and the supporting online measurement study guide. In no particular order, Stephen Neale, Richard Wise and Dr Stephanie Barnet for their technical guidance, suggestions and diplomacy in making things right that were clearly wrong. A special 'thank you' to Linda Packer and Alexandra Tisson for final proofreading and general encouragement through times when completion and delivery seemed unlikely.

Abbreviations

For abbreviations typically used when booking dimensions, see Chapter 2.08 and Appendix II
For abbreviations commonly adopted for units of measurement, see Chapter 4.03.02

4D BIM	Time: construction planning and management
5D BIM	Cost: quantity generation and construction costing
6D BIM	Client handover model: post-delivery facilities management
BCIS	Building Cost Information Service
BIM	see below for Level 0 BIM, Level 1 BIM, Level 2 BIM and Level 3 BIM
BQ	Bill of Quantity /Bills of Quantities
BS	Building & Construction Standards (see BSI)
BS EN	British Standard Eurocodes (replacing BS codes)
BSI	British Standards Institution
CAD	computer-aided Design
CAWS	Common Arrangement of Work Sections (not applicable to NRM2)
CDE	common data environment
CIBSE	Chartered Institute of Building Services Engineers
CIC	Construction Industry Council
CIOB	Chartered Institute of Building
COBie	Construction Operations Building Information Exchange
EDI	electronic data exchange
iBIM	integrated BIM (see Level 3 BIM below)
IEE	Institution of Electrical Engineers
IFC	Industry Foundation Class (protocol of common file formats such as)
JCT	The Joint Contracts Tribunal
Level 0 BIM	effectively means no collaboration
Level 1 BIM	limited collaboration
Level 2 BIM	collaboration between all parties but not necessarily a shared model
Level 3 BIM	assumes full collaboration and a shared project model located in a centralised repository
NBS	National Building Specification
NRM1	Order of cost estimating and cost planning for capital building works
NRM2	Detailed measurement for building works
NRM3	Order of cost estimating and cost planning for building maintenance works
PC	prime cost

PC Sum	prime cost sum
Prov Sum	provisional sum
RIBA	Royal Institute of British Architects
RICS	Royal Institution of Chartered Surveyors
SMM	Standard Method of Measurement
SMM7	Standard Method of Measurement, 7th edn
WS	Work Section (NRM2)

1   Introduction

1.01 Getting started

Before starting we need to establish the purpose and intention of measurement. If you were asked to measure a building, or the room you are in right now, most people would find a measuring tape, a clipboard and something to write with. You might sketch out the plan shape of the building or room and write the individual lengths and widths down as you go along.

> **ACTIVITY:** If you can, try sketching the plan shape of the room you are in right now. Get a tape measure and record the principal lengths and widths and annotate your sketch of the floor layout (plan) by recording the plan dimensions. It doesn't matter if you haven't got a tape measure to hand – you can simply 'pace the room'. This will obviously depend on how big your stride is so, for the purposes of getting you started we will just assume that one of your strides is equivalent to one metre.

TIP: most rooms are a regular shape (rectangular), so you will only need to measure one width and one length. Older buildings (and some poorly constructed new ones) may be 'out of square', so the above technique may not be appropriate. One way of checking whether a room is square is to measure both of its diagonals (from one corner of the room to the other). If the two diagonal dimensions are the same then it is a safe bet that the room is square and each of the corners is at 90 degrees.

Being able to read and draw to scale is an important part of what quantity surveyors call measurement.

Once you know the length, area or volume of the various components that go together to make up a finished building, you are in a position to allocate costs based on the units that you have measured. For example, if you were measuring a skirting board and you knew how much it cost per metre, finding the total cost once you know the internal room dimensions (perimeter length) is straightforward (Figure 1.1).

Recording dimensions from finished buildings is one way of finding out how much the building will cost. Once you are able to 'read' scale drawings and 'record dimensions' it becomes possible to cost buildings even before they are built. In essence, this is the service that a quantity surveyor offers to a client.

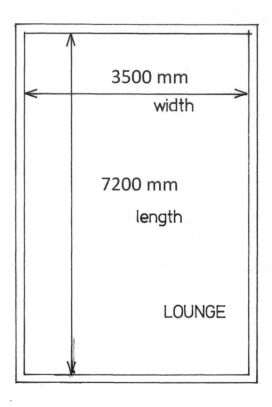

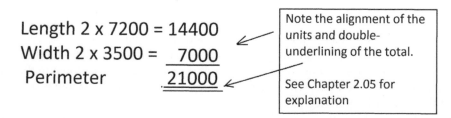

Figure 1.1 Dimensioned plan of room.

One of the first things that anyone thinking about commissioning a new building wants to know is 'how much?' By measuring the individual components and then allocating a cost to these, the likely finished build cost can be established. Simply adding together all of these individual components will give the total building cost.

OK, so this sounds simple enough but there are some other things to consider, especially when we are dealing with an idea that was 'in someone's head' and is now in the form of a drawing rather than a finished building.

With a finished building we can see its quality (or not) both in terms of the materials used and the workmanship. If we are measuring from a drawing this isn't going to be so obvious. This is of course significant when we are applying costs to our measured items. Consider the skirting we measured earlier around the perimeter of a room. A softwood skirting (pine) is likely to cost less than a hardwood (oak) equivalent; so we need to have some indication of the quality of the materials and the standard of workmanship required before we can state with certainty what things will cost.

So a set of measurements from a scale drawing isn't enough on its own. In order to cost a building we need to know two things: how much is there (the quantity) and what quality of finish/standard is required/expected (the specification).

1.02 Context and purpose

To the uninitiated, the phrase 'building measurement' brings to mind the fixed notion of a tape, a building and some numbers. One thing it probably does not immediately imply is cost and forecasting. Yet the purpose of measurement in this context is inextricably linked with providing an assessment of the cost of a building long before work has commenced on site. Initially, and most importantly, someone requires the provision of a new building. Normally they are likely to approach an architect so that their ideas and intentions can be set down on paper. It is very likely, even at this early stage, that they will need to know how much the building design proposals are going to cost (Figure 1.2a, b).

Armed with a set of drawings, a scale rule and a calculator, measurements can be taken from these drawings and a document produced. This document identifies in some detail the component parts of the proposed works, together with their quantity, and will allow construction costs to be allocated to the appropriate parts of the building. Having costed each component, a forecast for the scheme can be established (Figure 1.3).

Figure 1.2 A building is proposed – how much will it cost?

Figure 1.3 A forecast of the proposed scheme is prepared.

To enable this forecast of cost to be made with any confidence, a number of basic principles must be in place. It is very important that none of the building operations are overlooked and that the items which have to be costed are presented in a recognisable form. The consequence of an error in measurement or ambiguity in a description could result in the client being ill-advised with regard to the eventual cost of building operations.

1.03 General principles

Having identified the purpose of measuring building work it is necessary to establish the general principles that will ultimately result in a document which is mutually understood and conveys the scale and extent of the construction. A consistent approach is necessary both in terms of presenting this finished document and setting down dimensions so that others are able to understand our approach (Figure 1.4).

It is not hard to imagine the confusion that would result if everyone adopted their own set of rules when measuring building work. It was exactly this situation that prompted the publication in 1922 of the first nationally recognised (UK) set of rules for the measurement of building work. An indication of the chaotic state of affairs that prevailed prior to this

Figure 1.4 Clearly understood set of documents.

publication can be gleaned from reading the preface of this very first edition. Phrases such as 'diversity of practice' and 'idiosyncrasies of individual surveyors' suggest a picture of confusion and doubt for the hapless early twentieth-century contractor. Almost a century later, the same set of principles still apply. Now in an eighth edition, the current document *NRM2: Detailed Measurement for Building Works* provides the basic principles for the measurement of building work.

A report commissioned by the RICS Quantity Surveying and Construction Professional Group (Measurement-based procurement of buildings; 2003) claimed that the (then) current version of the Standard Method of Measurement (SMM7) was out of date and represented a time when bills of quantities and tender documents were required to be measured in greater detail than was warranted by procurement practice. At the same time it reported that the rise in the use of design and build procurement had encouraged the use of contractors' bills of quantities where few documents were prepared in a standard recognised form.

Importantly the report confirmed that some form of measurement remained necessary in the procurement of buildings, and that any new method of measurement would need to be flexible enough to accommodate the different ways that measurement was used. So any replacement document would need to be accessible and functional for a variety of different clients, contractors, subcontractors and suppliers in the procurement process. In short, it was time for an overhaul and an update.

Rather than 'reinventing the wheel' each time it was necessary to update construction costs, the new approach would need to allow the measured information and associated cost data to be passed on to the next stage. Different approaches to measurement and quantification at different stages in the design would of course remain necessary. However, wherever possible, any measurement/costs established for early-stage cost advice could be refined to inform approximate estimates, cost plans, bills of quantities and eventually the maintenance and repair costs of an operational building. This was the philosophy that underpinned the drafting of a suite of three interoperable documents – NRM1, NRM2 and NRM3 (Figure 1.5).

NRM1: Order of cost estimating and cost planning for capital building works.
NRM2: Detailed measurement for building works.
NRM3: Order of cost estimating and cost planning for building maintenance works.

Measurement can therefore be identified as the starting point from which construction costs are established. There is a standard format for the presentation of measured work and a set

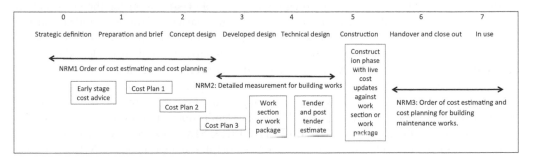

Figure 1.5 Timeline showing the NRM suite of documents with stage of estimating/costing mapped against RIBA Plan of Work (2013).

of rules that are mutually known and accepted. These rules are embodied in a document called *The New Rules of Measurement 2* (NRM2). *The New Rules of Measurement 2* is generally recognised by the acronym NRM2 and further defined by the phrase 'detailed measurement for building work'.

In turn, measurement provides the basis for the preparation of a Bill of Quantities (BQ). This document sets out the quality and quantity of all the component parts necessary for the construction of the works. It is prepared in a predetermined order which, in normal circumstances, would follow the same sequence as the work sections presented in NRM2.

Assuming a contract based on traditional tendering, each contractor would receive an identical set of tender documents including a BQ, architect's and structural engineer's drawings and a specification. Until recently these documents were prepared and provided to tenderers as hard copy, but more recently they became available electronically. See sections 4.01 and 4.02 for further details of Building Information Modelling (BIM) and e-tendering.

The costing columns of the BQs are completed, extended and totalled independently by each tendering contractor in order to establish a tender price for the completion of the construction work. Individual tenders are submitted and compared to the other tenders received. The client, acting on advice, normally accepts the most suitable tender and enters into a formal agreement with the selected contractor. The BQs, along with the drawings, specification and schedules, will eventually become part of the contract documentation to which both parties are formally contracted. While this costing activity achieves the principal function of the BQ (and thereby the measurement process), it also provides a valuable insight into the financial management of the project during, and to some extent after, the construction process. There are more sophisticated variants of the tender procedure described above but whichever technique is used, all construction costs will, at some stage, have been prepared from quantities established by the process of measurement.

The remainder of this text offers an interpretation by description and example of the techniques that are commonly practised in the preparation of detailed measurement in order to produce a BQ prepared in accordance with NRM2. Custom and practice will vary between individual surveyors, and the examples and procedures given here should not be considered as irrevocable. In spite of the rigours of NRM2, it is not unusual for individual practice to vary. With this in mind, the reader is advised that the techniques described and illustrated here should be regarded as recommendation rather than prescription.

1.04 The classification system

In the very first place, someone has to agree a system of classification for the construction process. It has to be sufficiently robust to embrace the variety of trades employed in the construction process, detailed enough to allow for technical distinctions and commonly understood by all those who use it. NRM2 was established to achieve these goals and has been adopted as the framework with which the detailed measurement of building work should be drafted. The classification system is loosely based on the pattern of trades employed during building operations. The order in which these are presented generally reflects the sequence of events as they are likely to occur on site. While the details might not always be obvious, the general coverage of each of the TABULATED WORK SECTIONS is clearly recognisable from the Work Section title. At this point it would be helpful to have a copy of NRM2 to hand so that the readers can begin to familiarise themselves with the structure and sequence adopted when preparing detailed measurement for building work. A free pdf download of NRM2 is available from the Royal Institution of Chartered Surveyors (RICS) – available to members and student members. In order for detailed measurement to take place it is a condition of NRM2 that sufficient design and production data are available (Figure 1.6)

For the purposes of measurement, NRM2 identifies a 'trade-specific' set of rules under each of the above Work Sections. The rules of measurement for the various trades or work

No.	Work Section:
2	Off-site manufactured materials, components and buildings;
3	Demolitions;
4	Alterations, repairs and conservation;
5	Excavating and filling;
6	Ground remediation and soil stabilisation;
7	Piling;
8	Underpinning;
9	Diaphragm walls and embedded retaining walls;
10	Crib walls, gabions and reinforced earth;
11	In-situ concrete works;
12	Precast/composite concrete;
13	Precast concrete;
14	Masonry;
15	Structural metalwork;
16	Carpentry;
17	Sheet roof coverings;
18	Tile and slate roof and wall coverings;
19	Waterproofing;
20	Proprietary linings and partitions;
21	Cladding and covering;
22	General joinery;
23	Windows, screens and lights;
24	Doors, shutters and hatches;
25	Stairs, walkways and balustrades;
26	Metalwork;
27	Glazing;
28	Floor, wall, ceiling and roof finishings;
29	Decoration;
30	Suspended ceilings;
31	Insulation, fire stopping and fire protection;
32	Furniture, fittings and equipment;
33	Drainage above ground;
34	Drainage below ground;
35	Site works;
36	Fencing;
37	Soft landscaping;
38	Mechanical services;
39	Electrical services;
40	Transportation; and
41	Builder's work in connection with mechanical, electrical and transportation installations.

Figure 1.6 NRM2 work sections (RICS Publications).

packages are presented in a table using a numerical prefix to distinguish each work section. The coding/classification system adopted by the NRM suite of documents (NRM1, NRM2 and NRM3) is intended to be interchangeable, allowing measurement and costings to be mutually interoperable across all documents. Details relating to the application of this coding system are explained in NRM2 2.15.3.

Specific measurement rules relating to particular trades can be found in NRM2 under each Work Section (see Figure 1.6). This is supplemented by general measurement rules that precede this and apply to the measurement of all construction operations. These rules comprise a uniform basis for measuring, describing and billing building works. To this end they identify the standard of accuracy necessary for recording quantities (NRM2 3.3 Measurement Rules for building work), together with definitions and interpretations for written descriptions. The application of these rules to specific trades forms the basis of Chapters 5 to 15 of this text. Reference is also made to the way in which the tabulated rules should be implemented. The following is an extract from NRM2 work section 5 Excavation and filling, and is included here to illustrate this (Figure 1.7).

Figure 1.7 shows an example of the tabular arrangement for NRM2 Work Section 5 Excavation and Filling. The first set of columns lists descriptive features that are commonly encountered in building operations, labelled 'items of work to be measured' (in this instance, excavation). The second narrow column is reserved for the appropriate unit of measurement (in this instance, cubic metres). The third provides sub-groups into which each main group can be further subdivided (known as level one) and the fourth distinguishes still further subdivision (known as level two). The fifth column (level three) identifies particular features unique to certain situations (in this case only one option is offered). Where appropriate, more than one feature may be selected from this column. The final column provides comments, definitions and interpretations. Descriptions are generated by selecting an appropriate descriptive feature from each column working from left to right (see selected text in Figure 1.7).

Item or work to be measured	Unit	Level one	Level two	Level three	Notes, comments and glossary
5 Site preparation—cont.	nr	4 Remove specific items.	I Dimensioned description sufficient to identify size and location of each item.		I Any existing items on site not specifically designated to remain including all types of rubbish such as abandoned cars, fridges and the like. 2 This excludes all but the simplest of building structures whose demolition is covered in Work Section 3: Demolitions. 3 Removal of any associated foundations, fixings, supports, fastenings and the like is deemed included.
6 Excavation, commencing level stated if not original ground level	m³	I Bulk excavation.	I Not exceeding 2m deep. 2 Over 2m not exceeding 4m deep. 3 And thereafter in stages of 2m.	I Details of obstructions in ground to be stated.	I Bulk excavation includes excavating to reduce levels or to form basements, pools, ponds or the like. For clarity each type of excavation may be measured and described separately. 2 Obstructions will be piles, manholes and the like that must remain undisturbed.
	m³	2 Foundation excavation.			I Foundation excavation includes excavating for strip and pad foundations, pile caps and all other types of foundations. 2 For clarity each type of excavation may be measured and described separately.

Figure 1.7 Building descriptions by selecting pre-defined phraseology from NRM2 Work Sections (example shows Excavation and Filling) (RICS Publications).

The measurer should observe and maintain the wording available within each set of horizontal lines when 'building' each description.

1.05 Traditional tendering

There are many reasons for requiring building work to be measured but the principal purpose is to identify cost. In order to give a client some indication of the likely cost for the proposal, early-stage construction cost estimates can be prepared. These are based on the client's anticipated use of floor space. While useful in providing a range of construction cost (lowest to highest), these are not regarded as reliable since they are frequently prepared on incomplete detail and limited specification. At best they provide the client with an indication of the likely cost. As the design progresses, an improved level of detail becomes available and early-stage cost advice can be refined and improved to reflect this.

Once the design is complete the measurement process can commence and a BQ can be produced. The BQ forms the basis of the tender process and will be priced by a number of competing contractors. Each will price the quantities independently to arrive at a total cost for the work (Figure 1.8).

The pricing process, once completed and totalled, provides the basis for a returned tender. For the successful contractor (usually the lowest returned tender) the priced BQ is incorporated with other documents into a formal contract. Both contractor and client are then legally bound to perform their respective parts of this contract: the contractor to build and the client to pay the agreed sum upon satisfactory completion of the works.

St Joseph's Upper School – Phase 1				£	p
5 EXCAVATION & FILLING					
The work comprises the excavation and associated substructures for a detached residential property all in accordance with Structural Engineer's drawing reference ATR200289. The ground is assumed to consist of 150mm of topsoil overlaying clay and sandstone depths all as detailed in the Structural Engineer's borehole log schedule dated 04/11/2020. The site is substantially level with existing GL established at 100.000. Groundwater level was established at 4.60m (95.400) below existing GL on 04/11/2020. No over or underground services cross the site					
Preliminary site work;					
Boreholes to determine ground conditions					
A	100mm diameter, 18.00 metres maximum depth	6	nr	720.00	4,320	00
B	100mm diameter, 24.00 metres maximum depth	4	nr	800.00	3,200	00
Removing trees					
C	girth 500 -1500mm	8	nr	280.00	2,240	00
D	girth 1500 -3000mm	4	nr	558.00	2,232	00
Removing tree stumps					
E	girth 500 -1500mm, backfilled with excavated materials	8	nr	120.00	960	00
F	girth 1500 -3000mm, backfilled with excavated materials	4	nr	180.00	720	00
Site Clearance					
G	Clear site of all vegetation and other growth, dispose of off site	1154	m²	2.82	3,254	28
Page 3/45			To Collection	16,926	28

description and quantity columns completed by quantity surveyor   costing columns completed by contractor

Figure 1.8 BQ page showing completed pricing and page collection total.

2 Traditional approaches to booking dimensions and bill preparation

2.01 Introduction

In the first instance, quantities will need to be extracted from the drawings, together with an appropriate description. This process, known as 'booking dimensions' or 'taking-off quantities', involves the measurer in either reading or scaling dimensions from the drawings. There are two distinct parts to this. The first involves the recording of quantities and the second requires a written description to accompany the quantity. The sequence adopted by the measurer in the initial phase of booking dimensions bears little relation to the eventual order of the finished Bill of Quantities (BQ). This is because 'taking-off' has been devised in order to assist the measurer with the speed and accuracy of recording dimensions and largely follows the sequence of events as they will occur on site. At a later stage these same booked dimensions will be arranged in the sequence of NRM2. For example, when measuring a foundation trench the excavation, disposal, concrete work and masonry are all measured at the same time, regardless of the eventual location of these items in the finished BQ. This particular pattern of measured items is generally known as the 'group method', since it reflects a common set of dimensions shared by a number of different trades. The grouping of measured items when recording dimensions is not determined by their eventual position in the BQ, but by the fact that they share a common set of base dimensions.

The alternative to the group method of measurement is known as the 'northern method', or the trade-by-trade approach. As the name suggests, each item should be measured in trade sequence (or the order dictated by the standard method). This means scanning all the drawings to determine the order of measurement and gives little opportunity for bringing together items that share a common set of dimensions. This approach is often adopted for small projects, but it is inappropriate for larger, more complex work.

While there will always be a need to understand the mechanism by which a BQ is produced, the present generation of quantity surveyor (QS) is likely to be assisted in this task by a PC and specialist software. Increasingly the adoption of Building Information Modelling (BIM) in conjunction with BQ software (QS Pro) can facilitate the automatic generation of a BQ. Providing the design is prepared as a building model it will be interoperable and (by default) allow data exchange between all parties who are part of the design team, including the QS. Details of BIM technologies are considered in more detail later (see 4.02).

Regardless of the opportunities that these new approaches offer (and they may be significant), it is important to recognise that a thorough grasp of underlying principles needs to be in place before the full utility of information modelling systems can be realised. To this end, the remainder of this chapter assumes a conventional approach using the group method of measurement.

Once the take-off is complete the measured items will be squared and totalled. All of the arithmetic should be checked. At this point the sorting process to establish the eventual order of the measured items (*abstracting*) can commence (see Section 2.11).

2.02 The principles of setting down dimensions

It is important to appreciate that the techniques adopted in the preparation of a BQ will vary from one individual to another and between one office and the next. For the most part it is assumed that the operation of transferring drawn and other design information into the descriptions and quantities that are collectively termed 'a Bill of Quantities' will be carried out without the use of a computer. Clearly there are significant advantages in using a computer package that can generate digitised quantities and automatically sort measured items into the appropriate Work Section. Nevertheless, in order to fully understand the detail of this process, it will be necessary to examine each step in turn and this cannot be fully appreciated by simply pressing a button. What follows is based on a procedure that has evolved over a number of years and is best described by the term 'traditional bill preparation' (Figure 2.1).

Dimensions are taken from a drawing and recorded on specially lined paper known as *dimension paper* (Figure 2.2)

2.02.01 Dimension column

Dimensions are read directly from the drawings and recorded to two decimal places of a metre in the middle of the three narrower vertical columns labelled *dimension column*. Alternatively, these dimensions might be scaled directly from the drawing and entered in the same way. Once

Stages in traditional BQ preparation

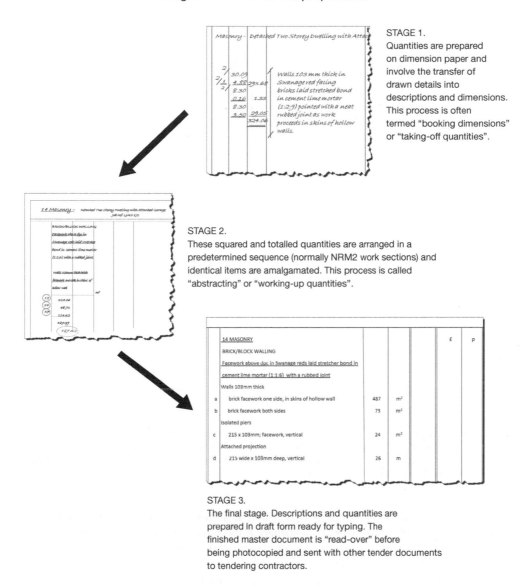

STAGE 1.
Quantities are prepared on dimension paper and involve the transfer of drawn details into descriptions and dimensions. This process is often termed "booking dimensions" or "taking-off quantities".

STAGE 2.
These squared and totalled quantities are arranged in a predetermined sequence (normally NRM2 work sections) and identical items are amalgamated. This process is called "abstracting" or "working-up quantities".

STAGE 3.
The final stage. Descriptions and quantities are prepared in draft form ready for typing. The finished master document is "read-over" before being photocopied and sent with other tender documents to tendering contractors.

Figure 2.1 Stages in traditional BQ preparation.

these have been recorded or 'booked', it will be necessary to provide some form of description. The largest of the four columns, labelled the *description column*, is used for this purpose.

Before describing the purpose of the remaining two columns, it would be prudent to spend some time looking at the various ways in which dimensions are recorded. The technical term for entering dimensions in this way is 'booking dimensions', and these should always be recorded to *two decimal places of one metre* (NRM2 3.3.2.d). Accordingly, 3067mm would be

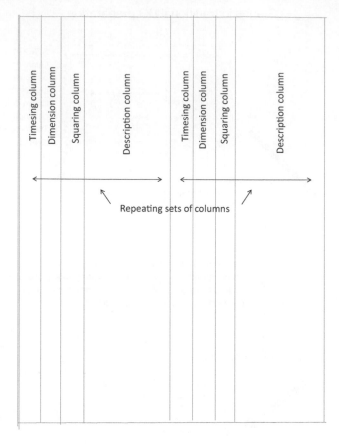

Figure 2.2 Dimension paper. An A4 page is divided vertically into two identical halves, each comprising a set of four columns. These are labelled above for the purpose of identification (Figure 2.2). The extra column on the extreme left is termed the binding margin and would not normally be used for recording dimensions.

recorded in the dimension column as 3.07. There is no need to write the word metre or use the letter m, since all dimensions are recorded in the same way. Even when the dimension is a whole unit (5 metres), two zeros should be used after the decimal place (it should be recorded in the dimension column as 5.00).

The following shows examples of dimensions as they might appear on an architect's drawing, together with the corresponding dimensions as they should appear once they have been booked on the sheet of dimension paper (Figure 2.3).

So far we have only considered recording measurable items in the dimension column: as an example, items represented by the unit of length include rainwater guttering, drainage pipe runs, rafters and skirtings. It would be inappropriate to use this same unit for items such as excavation, which not only have a length but in addition, a width and a depth. It may have been noted in the previous examples that a line was drawn across the dimension column under each set of recorded dimensions. This identified each single entry as an individual length. The technical term for this unit of measurement is *linear metres*. The principal unit of

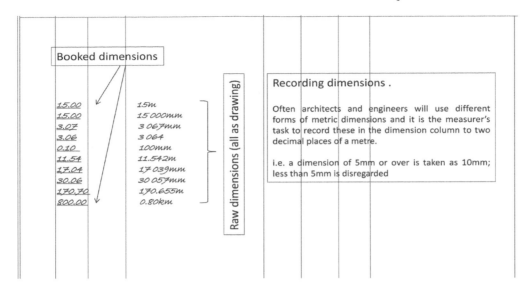

Figure 2.3 Recording dimensions.

measurement associated with excavation or in situ concrete is *cubic* metres, and this is shown in the dimension column with *all three dimensions* set down one above the other and presented as a dimension set by the inclusion of a horizontal line across the dimension column under the last of these three dimensions. In similar fashion, dimensions grouped together in pairs are automatically associated with items that have been measured in *square* metres, such as brickwork and plasterwork. In some cases it is difficult to identify an appropriate unit of measurement and in such instances counting or *enumeration* is used. Door furniture, manhole covers and sanitary appliances are all recorded in the dimension column using enumeration. These appear as whole numbers in the dimension column, with a line drawn horizontally beneath each single entry (there is no decimal point required with enumerated items).

To summarise:

- Dimensions are recorded to two decimal places of a metre.
- There is no need to identify the unit of measurement since this is clear from the presentation.
- Where dimensions appear in sets of two or three, these will be multiplied together (squared) to show an area or volume.
- Clear, legible presentation, in ink, with items well spaced.
- The order of recording dimensions is
  - length (followed by)
  - width or breadth (followed by)
  - vertical height or depth.

This last point is of no significance in regard to items of length or enumeration, but is of great value when the need arises to trace the build-up of dimensions for areas and volumes (see Figure 2.8).

The following are examples of each of these.

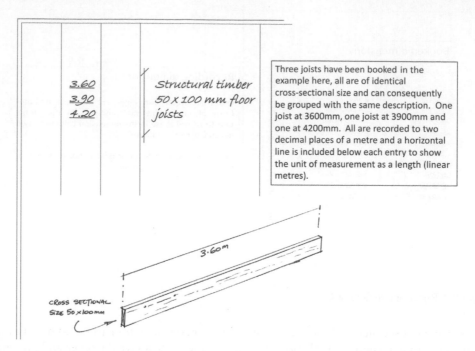

3.60
3.90
4.20

Structural timber
50 x 100 mm floor
joists

Three joists have been booked in the example here, all are of identical cross-sectional size and can consequently be grouped with the same description. One joist at 3600mm, one joist at 3900mm and one at 4200mm. All are recorded to two decimal places of a metre and a horizontal line is included below each entry to show the unit of measurement as a length (linear metres).

Figure 2.4 Booking linear dimensions (length).

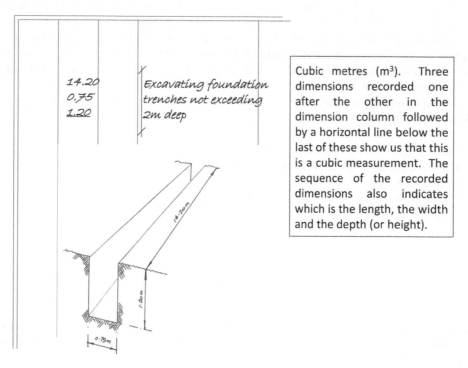

14.20
0.75
1.20

Excavating foundation
trenches not exceeding
2m deep

Cubic metres (m^3). Three dimensions recorded one after the other in the dimension column followed by a horizontal line below the last of these show us that this is a cubic measurement. The sequence of the recorded dimensions also indicates which is the length, the width and the depth (or height).

Figure 2.5 Volume (m3).

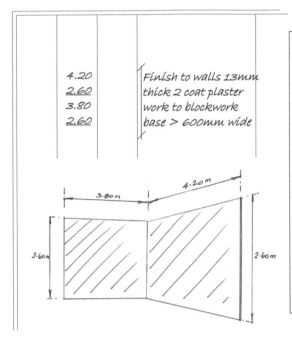

4.20
2.60
3.80
2.60

Finish to walls 13mm thick 2 coat plaster work to blockwork base > 600mm wide

4.20m

3.80m

2.60m

2.60m

Square metres (m²) The sketch shows an internal isometric view of a room but only two of the four walls are shown. In this instance both walls are plastered and share the same specification together with a floor to ceiling height of 2.60 metres. One is 4.20m long and the other is 3.80m long. We can tell that these are measured in square metres because there are two dimensions followed by a horizontal line. In both cases the first dimension is the length and the second the height.

Figure 2.6 Area (m²).

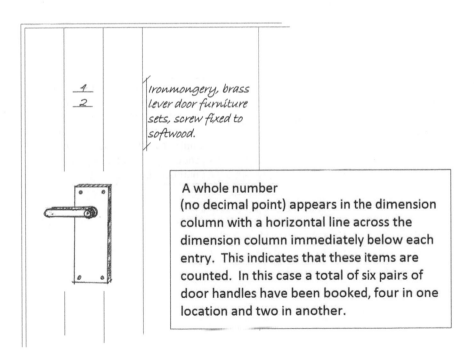

1
2

Ironmongery, brass lever door furniture sets, screw fixed to softwood.

A whole number
(no decimal point) appears in the dimension column with a horizontal line across the dimension column immediately below each entry. This indicates that these items are counted. In this case a total of six pairs of door handles have been booked, four in one location and two in another.

Figure 2.7 Enumerated items (nr).

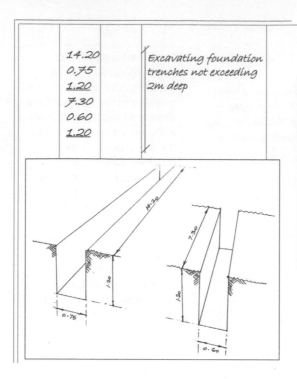

Here there are two trenches of differing lengths and widths but with an identical depth. Each requires its own set of dimensions and follows the convention of three dimensions followed by a horizontal line to indicate volume. Dimensions are presented in the sequence of length, width then depth.

Figure 2.8 Entering multiple quantities.

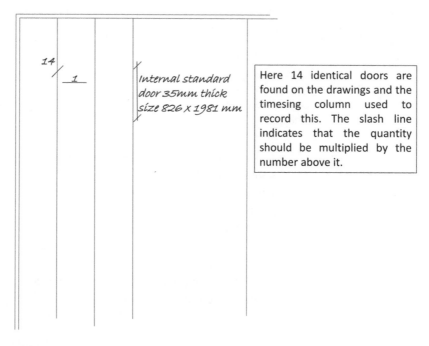

Here 14 identical doors are found on the drawings and the timesing column used to record this. The slash line indicates that the quantity should be multiplied by the number above it.

Figure 2.9 Timesing.

2.02.02 *The timesing column*

So far we have only considered using the dimension column and the description column.

The term 'timesing' (Figure 2.9 and Figure 2.10) is used by measurers where there are a number of identical sets or repeats of the same item being measured. For example, when measuring a number of identical internal doors it is convenient to simply record the dimension and description once, and then 'times' this by however many times the item being measured occurs. In order to do this the measurer uses the furthest left of the three narrow columns and inserts the number of times the item in question is repeated. This is followed by inserting a short diagonal 'slash' line between the booked dimension and the number by which this will be 'timesed'. It may also go some way towards providing an indication of the layout from the presentation of the dimensions (see Figure 2.27, adjustments).

2.02.03 *The squaring column*

Attention can now turn to the function of the squaring column. As previously noted, dimension figures are recorded in the dimension column to indicate the unit of measurement. In the case of area and volume, dimensions will be multiplied and the result of this computation entered in the squaring column. There is never any need to include a multiplication sign, since this is assumed (Figure 2.11).

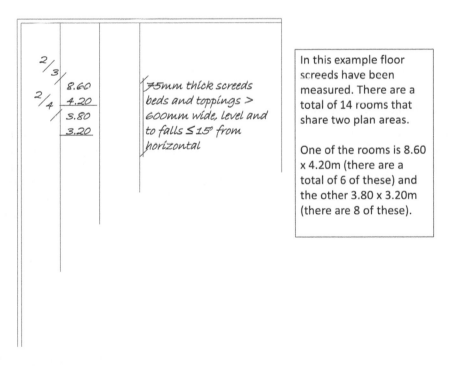

In this example floor screeds have been measured. There are a total of 14 rooms that share two plan areas.

One of the rooms is 8.60 x 4.20m (there are a total of 6 of these) and the other 3.80 x 3.20m (there are 8 of these).

Figure 2.10 Timesing.

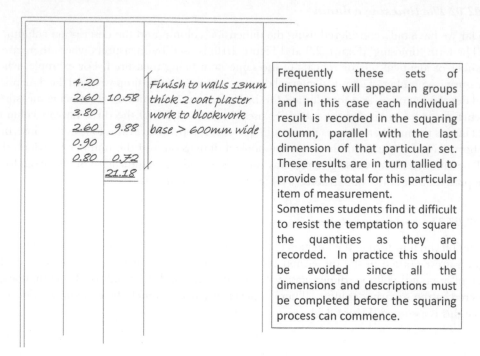

Figure 2.11 Squaring dimensions.

Past experience has shown that students find it difficult to resist the temptation to square the quantities as they are recorded. In practice this should be avoided since all the dimensions and descriptions must be completed before the squaring process can commence.

2.02.04 *The description column*

As the title suggests, this is where the worded part of the measurement is recorded. The order and form of wording is important, as this should convey concisely all the information necessary to allow a price to be established. The tabulated arrangement of NRM2 provides an appropriate framework around which to build descriptions. Despite this, there still remains a degree of flexibility in the interpretation and presentation of written descriptions. As far as possible, the structure and terminology of NRM2 has been adopted when writing descriptions in the preparation of this text.

This is one of the more difficult parts of booking dimensions, and those new to drafting descriptions are advised to spend some time looking through a BQ before attempting to write descriptions for measured work (Figure 2.12).

2.03 Dotting-on

Where identical dimensions are repeated it may be necessary to add rather than to multiply. For example, six pits of identical size are to be excavated and the following dimensions are recorded (Figure 2.13a).

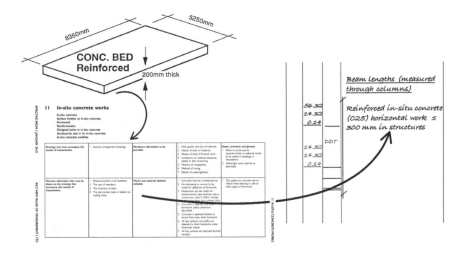

Figure 2.12 Interpreting NRM2: relationship between drawings, NRM2 clauses and booked dimensions.

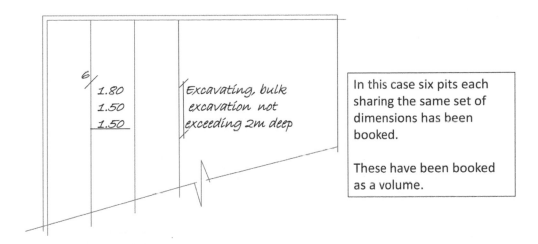

Figure 2.13a Timesing.

Subsequently, an additional two pits are spotted. Rather than enter another set of dimensions together with an identical description, the original recorded dimensions can be adapted by 'dotting-on' another two as follows (Figure 2.13b).

This technique may also be carried out in the timesing column and (technically) any number of additions can be made, although this will be somewhat restricted by the limitations of space and the need for legible presentation. The dot will usually be placed below the original figure to allow more space, and the timesing slash will be placed above it. In practice, these two techniques are often combined (see Figures 2.14 and 2.15).

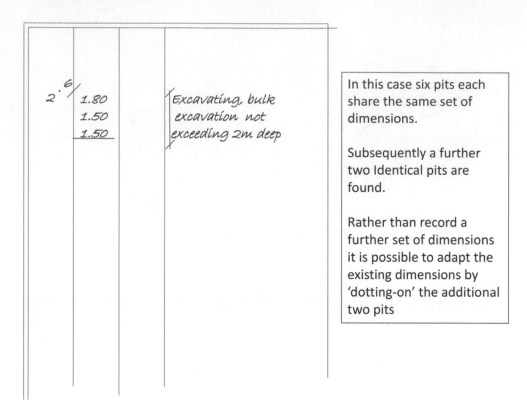

Figure 2.13b Dotting-on.

In this case six pits each share the same set of dimensions.

Subsequently a further two Identical pits are found.

Rather than record a further set of dimensions it is possible to adapt the existing dimensions by 'dotting-on' the additional two pits

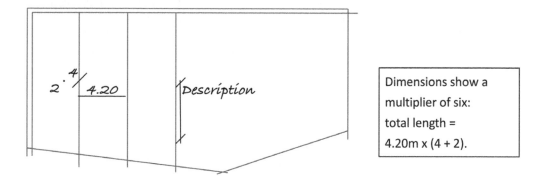

Figure 2.14 Combining timesing and dotting-on.

Dimensions show a multiplier of six:
total length =
4.20m x (4 + 2).

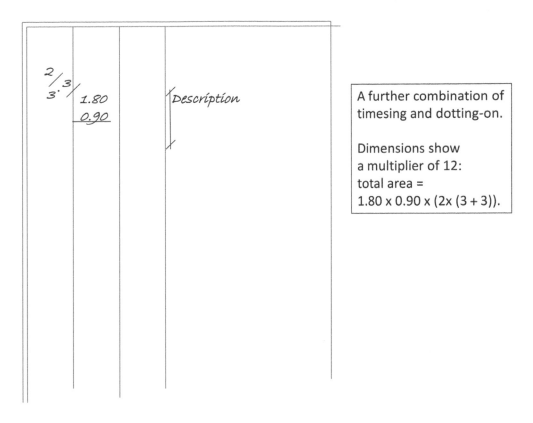

A further combination of timesing and dotting-on.

Dimensions show
a multiplier of 12:
total area =
1.80 x 0.90 x (2x (3 + 3)).

Figure 2.15 Dotting-on and timesing.

2.04 Geometric forms

Frequently it is necessary to book dimensions for triangles, circles and other irregular figures. The important point to remember when booking dimensions using geometric formulae is the unit of measurement. Most geometric formulae are quite easily transposed to booked dimensions. For example, the formula for the area of a triangle is half the base multiplied by the height. This can be recorded on dimension paper as follows (Figure 2.16).

The area of a circle with radius 8.64 m (formula πr^2) can be recorded as seen in Figure 2.17, and of a sector of a circle as in Figure 2.18.

If, on the other hand, we needed to record the circumference (perimeter length) of this circle, only a single dimension would need to appear in the dimension column (Figure 2.19; formula $2\pi r$).

The inclusion of a fraction in combination with the above can give the circumference length of any portion of the circle, e.g. a semi-circle (see Figure 2.19).

The formula for regular geometric forms and how these should appear as booked dimensions is included in Appendix II.

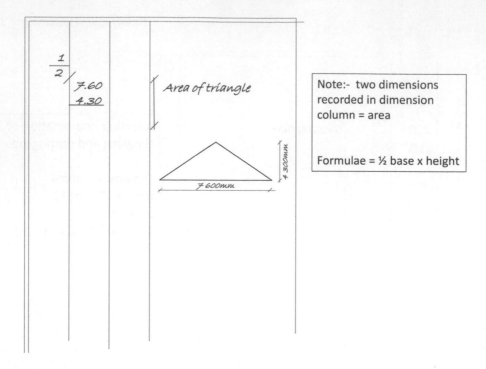

Figure 2.16 Geometric forms area of triangle.

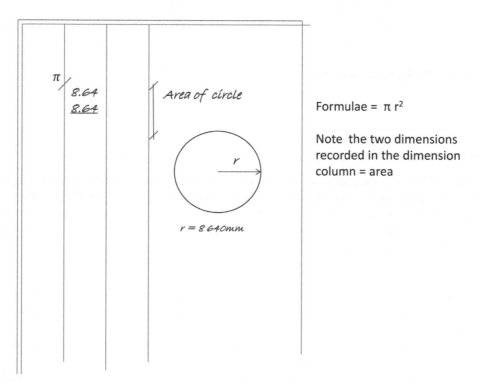

Figure 2.17 Geometric forms area of a circle.

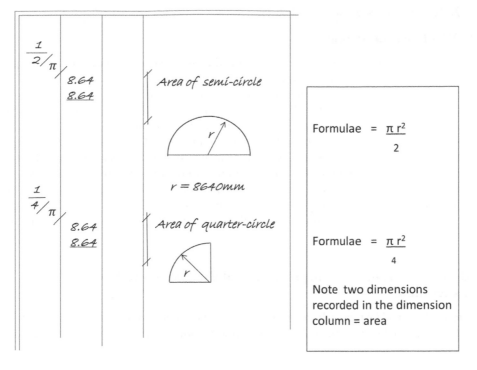

Figure 2.18 Geometric forms area of semi and quarter circle.

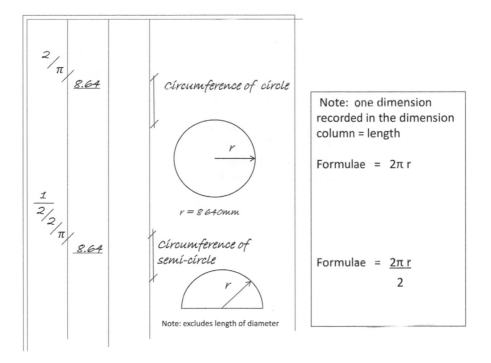

Figure 2.19 Geometric forms perimeter length of circle and semi-circle.

2.05 Waste calculations

Ideally, dimensions can be read directly from the drawings and entered to two decimal places of a metre on dimension paper. Frequently this direct transfer is not possible since some adjustment is required to the dimensions before they can be booked. These preliminary calculations are known as 'waste calculations' or 'side-casts' and are presented to the nearest millimetre (three decimal places of a metre) on the right-hand side of the description column, immediately above the item to which they relate (Figure 2.20). Once the required dimension is identified by waste calculation, it is reduced to two decimal places after transfer to the dimension column. There may be a temptation to scribble these in note form or even carry out simple arithmetic in the head. Both should be avoided since it is important to identify and demonstrate to others the process by which the dimension was established. Double-underlining in a waste calculation usually indicates that the result has been transferred to the dimension column.

2.06 Bracket lines, the ampersand and the group method of measurement

Inevitably there will be cases where more than one set of dimensions relates to a single description. These can be linked to the recorded dimensions with a *bracket line* (Figures 2.20 and 2.21).

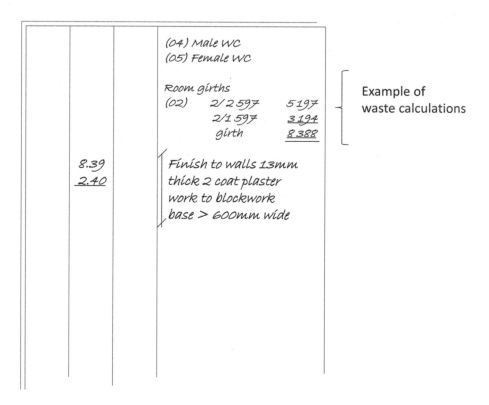

Figure 2.20 Waste calculations.

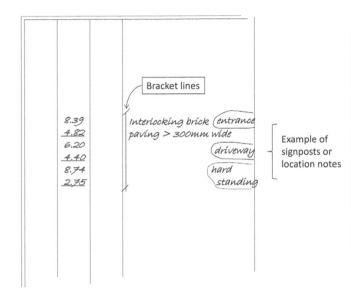

Where two or more descriptions apply to a single set of dimensions each description is separated by an ampersand. This process of 'anding-on' is fundamental to the group method of measurement and avoids the need to write down dimensions more than once. Care should be taken to ensure that the same unit of measurement applies to each of the items linked together by an ampersand, as the total of all the quantities on the left of the bracket, relate to EACH and EVERY description on the right of the bracket.

Figure 2.21 Bracket line and signposting.

Where two or more descriptions apply to a single set of dimensions, each description is separated by an ampersand. This process of 'anding-on' is fundamental to the group method of measurement and avoids the need to write down dimensions more than once. Care should be taken to ensure that the same unit of measurement applies to each of the items linked together by an ampersand, as the total of all the quantities to the left of the bracket relates to *each and every* description to the right of the bracket.

On occasion, it is convenient to use an ampersand to link two or more items that have different units of measurement. This is perhaps best demonstrated by Figure 2.23.

In this case, three different classes of work have been grouped together as they share a common set of base dimensions (in this instance it is the girth or perimeter length of a room). The first three items are measured in linear metres while the last two are measured as an area. Rather than starting a new set of dimensions, it is far simpler to introduce a conversion factor (in this case the constant floor-to-ceiling height). This must be made clear to the person carrying out the squaring and is usually shown by inserting a horizontal arrow across the dimension and squaring column into the description column where the conversion takes place (in this one instance a mathematical times symbol [×] can be used). A space for the resulting quantity must be left, and the appropriate unit of measurement should be identified after the space.

This is a more advanced technique, which anyone new to recording dimensions is advised to avoid until such time as the basic principles of booking dimensions are clearly understood.

2.07 Spacing of dimensions and signposting

The spacing of dimensions and descriptions is an important part of the measurement process. Clear, well-spaced dimensions are easy to follow and can be readily checked by others. While

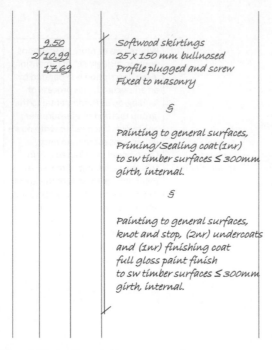

Where two or more descriptions apply to a single set of dimensions each description is separated by an ampersand. This process of 'anding-on' is fundamental to the group method of measurement and avoids the need to write down dimensions more than once. Care should be taken to ensure that the same unit of measurement applies to each of the items linked together by an ampersand, as the total of all the quantities on the left of the bracket relate to EACH and EVERY description on the right of the bracket.

Figure 2.22 Anding-on (the ampersand).

there are no written rules about the presentation of measured items, Figure 2.24 provides an indication of the spacing and layout of a typical sheet of booked dimensions.

Every effort should be made in the take-off to ensure that dimensions and waste calculations can be traced back to the drawing. Signposts or location notes can be used in the description column to provide this cross-reference. These should appear on the right of the description column parallel with the dimension to which they relate. A simple line or ring around this location note prevents it being read as part of the description (Figure 2.21).

2.08 Abbreviations

The use of abbreviations when writing descriptions is commonly practised by measurers. Apart from the practical limitations of space, a great deal of time is saved by shortening the more frequently used words. A full list of the more commonly recognised abbreviations is given in Appendix I. Individual surveyors and practices tend to develop their own forms of abbreviation and, although there are no hard and fast rules, it is always important to bear in mind that others must be able to fully understand the written description. While abbreviations are acceptable in both the take-off and the abstract, their place in the completed BQ should be restricted to those listed in NRM2 1.6.1. Should there be the slightest chance that a word or term could be misinterpreted in the finished BQ, it must be written in full.

Many of the descriptions in a take-off are repeated several times often in the space of a few pages. Where this occurs time can be saved by referring to a previous similar or identical description. This is achieved by using the abbreviation abd. (as before described). See Figure 2.25.

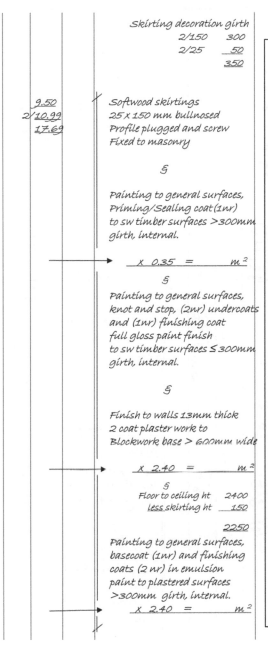

Skirting decoration girth
2/150	300
2/25	_50_
	_350_

9.50
2/10.99
_17.69_

Softwood skirtings
25 x 150 mm bullnosed
Profile plugged and screw
Fixed to masonry

&

Painting to general surfaces,
Priming/Sealing coat (1nr)
to sw timber surfaces >300mm
girth, internal.

x 0.35 =   m²

&

Painting to general surfaces,
knot and stop, (2nr) undercoats
and (1nr) finishing coat
full gloss paint finish
to sw timber surfaces ≤300mm
girth, internal.

&

Finish to walls 13mm thick
2 coat plaster work to
Blockwork base > 600mm wide

x 2.40 =   m²

&

Floor to ceiling ht	2400
less skirting ht	_150_
	_2250_

Painting to general surfaces,
basecoat (1nr) and finishing
coats (2 nr) in emulsion
paint to plastered surfaces
>300mm girth, internal.

x 2.40 =   m²

In this case three different classes of work have been grouped together as they share a common set of base dimensions (in this instance it is the girth or perimeter length of a room). The first and third items are measured in linear metres while the second and last two are measured as an area. Rather than starting a new set of dimensions it is far simpler to introduce a conversion factor (in this case the constant floor to ceiling height is included with the last two booked dimensions).

This must be made clear to the person carrying out the squaring and is usually shown by inserting a horizontal arrow across the dimension and squaring column into the description column where the conversion takes place (in this one instance a mathematical times symbol [x] can be used). A space for the resulting quantity must be left and the appropriate unit of measurement should be identified after the space.

Figure 2.23 Constant dimension.

STRUCTURAL WALLS 2

Ext. girth of brickwork

2/6950	13900
2/8300	16600
ext. girth	30 500

& ext. skin bwk.
less 4/2/½/103   412
                  30 088

30.09	Walls facework o/s half
4.88	brick thk (103) in Funtley
2/ 8.30	red fcg. bwk laid st. bond (gable end raisg)
2/½/ 0.16	in g.m. (1:2:9) pointed
8.30	with a neat nibbed jt (gable ends)
3.50	a.w.p.

Centre line of cavity

Ext. girth a.b.   30 500

less   103   4/2/
(5/2)   25   128   1 024
        128
                  29 476

29.48	Forming cavities in
4.88	hollow walls 50mm
2/ 8.30	wide inc. 4 nr. galv. (gable end raisg)
2/½/ 0.16	wall ties per m²
8.30	(gable ends)
3.50	

& int skin

ext. girth ab   30 500

less   103   4/2/
        50   228   1 824
(150/2) 75
        228
                  28 676

Blockwork ht to plate

dpc to ffloor	2400
floor joist	225
ffloor joist to clg.	2350
	4975

less   plate 50
        bed 10   60
                  4915

28.68	Walls 150mm insul.
4.92	blockwk laid stretch (gable raisg)
2/ 7.99	bond in g.m. (1:2:9)
2/⅓/ 0.16	(gable)
7.99	
3.50	

Adj. for attached garage

	Deduct	
½/		
	5.80	Walls facework o/s (attached garage gable)
	3.60	half bk thk (103) in
		Funtley red all a3d.

&

Add
Walls facework o/s
180mm thk solid conc.
blockwork (7N/mm²)
laid st. bond in
g.m. (1:2:9)

Figure 2.24 Sample layout of dimension paper showing booked dimensions, descriptions, waste calculations, ampersands and bracket lines.

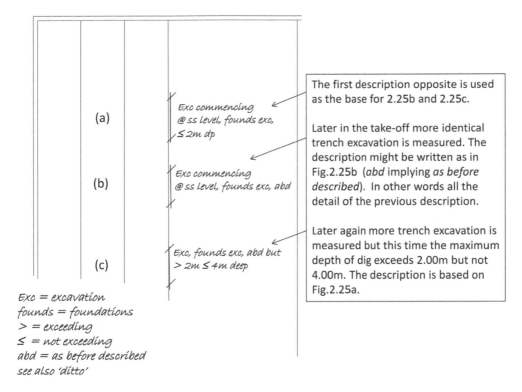

(a)

Exc commencing
@ ss level, founds exc,
≤ 2m dp

(b)

Exc commencing
@ ss level, founds exc, abd

(c)

Exc, founds exc, abd but
> 2m ≤ 4m deep

The first description opposite is used as the base for 2.25b and 2.25c.

Later in the take-off more identical trench excavation is measured. The description might be written as in Fig.2.25b (*abd* implying *as before described*). In other words all the detail of the previous description.

Later again more trench excavation is measured but this time the maximum depth of dig exceeds 2.00m but not 4.00m. The description is based on Fig.2.25a.

Exc = excavation
founds = foundations
> = exceeding
≤ = not exceeding
abd = as before described
see also 'ditto'

Figure 2.25 Abbreviations and short written descriptions. The description in (a) appears on the dimension sheet. Later in the take-off, more trench excavation is measured and the description might be written as (b) implying all the detail of the previous description. Later again more trench excavation is measured, but this time the maximum depth of dig exceeds 2.00 m but not 4.00 m and this is described in (c).

Note: For the sake of clarity, dimensions are not shown.

The word 'ditto' is often used in the same way, although in this instance it is understood to refer to the immediately previous description. For example, the following dimensions might be booked when measuring drainage work (Figure 2.26).

The full description for the second example should read, 'Drain runs, average trench depth 1000–2000 mm, 110 mm-diameter glazed clayware jointed with plastic couplings in the running length bedded and surrounded in pea gravel.'

Care should be taken when using either technique, since it is easy to misdirect the intended back-reference.

2.09 Adjustments

This is the term that measurers use to describe an alteration to a set of previously recorded dimensions. Generally this is necessary because it is easier and safer to over-measure an item in the first instance and make a deduction or adjustment later. Consider the measurement of a carpet for an office floor which is interrupted around its perimeter by the 'footprint' of a number of concrete columns. Rather than attempt to break this area down into a series of strips, it is far simpler to over-measure initially and then deduct the area of flooring covered by the columns.

It is usually assumed that all booked items are additions and there is consequently no need to write the word *add* each time a description is recorded. The opposite of this is the case with respect to deductions, and it is vital that this is made clear to the squarer. A deduction will generally follow the item which gave rise to that adjustment (see previous example).

There are two methods generally recognised for making adjustments, and the difference is purely a matter of presentation. An example for each of these follows.

Care should be taken when making adjustments to a number of descriptions linked to the same quantity using an ampersand. This can become particularly confusing unless the word 'deduct' is included for every adjustment and not just the first (Figure 2.28).

As shown in Figure 2.29, even though the intention of the measurer is clear, once the first description is transferred to the abstract and lined through, the squarer has lost all reference to the instruction to deduct the remaining items and it is likely that these quantities will be added rather than omitted.

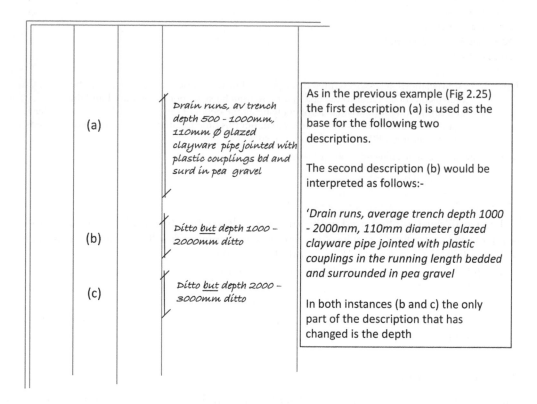

Figure 2.26 Short written descriptions and abbreviations.

Note: For the sake of clarity dimensions are not shown.

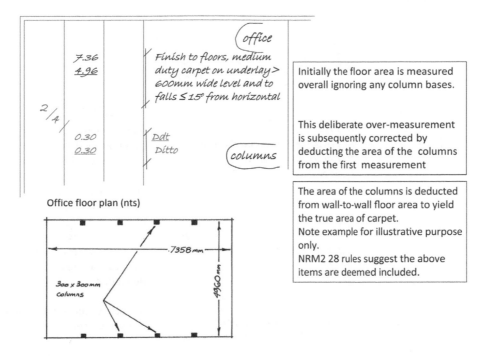

Office floor plan (nts)

Figure 2.27 Making adjustments.

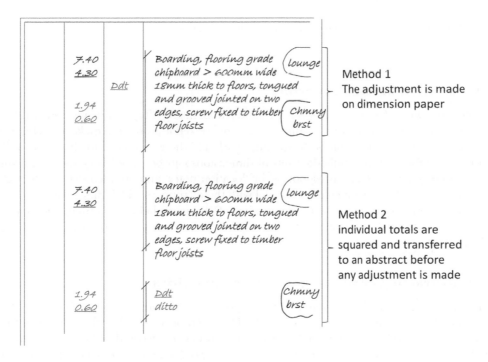

Figure 2.28 Options when making adjustments.

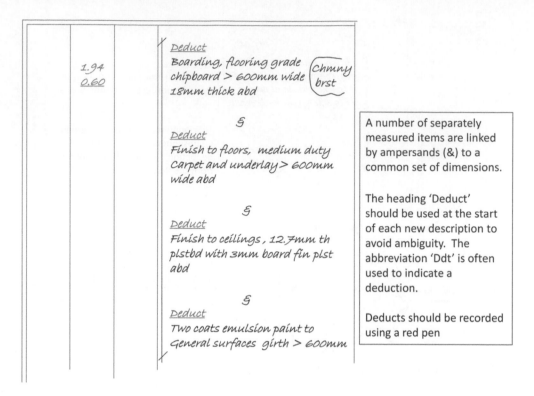

Figure 2.29 Adjustments (deductions) sharing the same dimensions.

2.10 Location notes/signposting

An often overlooked, but nonetheless important, part of taking-off quantities is the presentation of dimensions and descriptions. As dimensions are booked it is useful to make a brief note to identify the source or location of the quantities. Should there be a need to check how a BQ item was established, the individual sets of dimensions can be traced back to the drawing. These notes are referred to as 'signposts' or 'location notes' and are recorded on the right-hand side of the description column parallel with the quantity to which they relate. To avoid these becoming confused with the description of the work being measured, they can be distinguished from the wording of the description with a semi-circle or ring around them (Figure 2.30).

2.11 Abstracting or working-up bills of quantities

The sequence adopted by measurers using the group method of measurement largely follows construction operations as they occur on site. However, once the take-off is complete, these measured items will need to be collated, like items must be merged and deduction adjustments made. This process, known as *abstracting* or *working-up quantities*, is carried out on specially lined A3-sized paper (Figure 2.33).

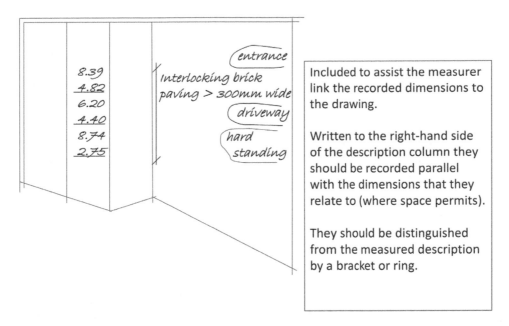

8.39	*entrance*	Included to assist the measurer link the recorded dimensions to the drawing.
4.82	*Interlocking brick*	
6.20	*paving > 300mm wide*	
4.40	*driveway*	Written to the right-hand side of the description column they should be recorded parallel with the dimensions that they relate to (where space permits).
8.74	*hard*	
2.75	*standing*	

They should be distinguished from the measured description by a bracket or ring.

Figure 2.30 Signposting (Location Notes).

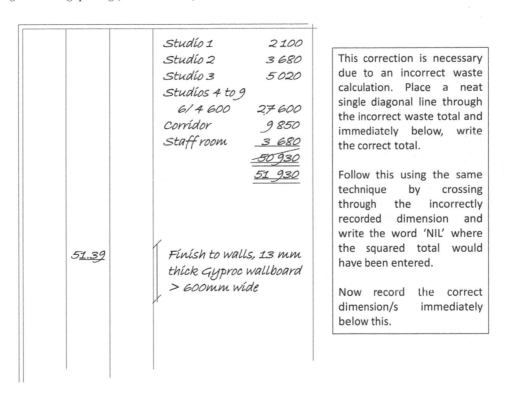

Studio 1	2 100
Studio 2	3 680
Studio 3	5 020
Studios 4 to 9	
6/ 4 600	27 600
Corridor	9 850
Staff room	3 680
	~~50 930~~
	51 930

51.39 — *Finish to walls, 13 mm thick Gyproc wallboard > 600mm wide*

This correction is necessary due to an incorrect waste calculation. Place a neat single diagonal line through the incorrect waste total and immediately below, write the correct total.

Follow this using the same technique by crossing through the incorrectly recorded dimension and write the word 'NIL' where the squared total would have been entered.

Now record the correct dimension/s immediately below this.

Figure 2.31 Correcting errors (1).

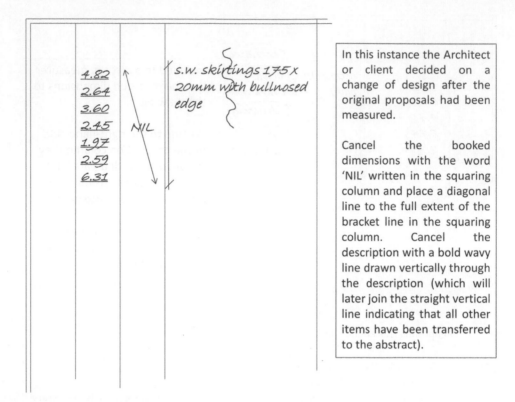

Figure 2.32 Correcting errors (2).

In the top banner at the head of each sheet of abstract paper, a Work Section heading is recorded together with any other references for the project. Each measured item is copied from the dimension column and transferred to the abstract in NRM2 Work Section order. As this transfer takes place, a vertical line is placed through the (now abstracted) description on the sheet of dimension paper. This is carried out in an effort to avoid either a double transfer, or the possibility of an item being overlooked (Figure 2.33).

On the abstract, the descriptions are copied spanning two columns. NRM2 headings/sub-headings are included to provide the framework for the final document. A horizontal line is drawn below each transferred description and the squared quantity entered below this line – additions on the left and deductions on the right. To provide a cross-reference to the dimension page, each squared quantity is labelled with the dimension page number in order to provide the source of the build-up dimensions. Finally, the appropriate unit of measurement, identified by the letters L (Linear), S (Square Area), C (Cubic Volume) or Nr. (Enumerated Items), should be entered. Once these items have been transferred, any reference to the unit of measurement as it was initially recorded on dimension paper would be lost.

Each transferred item is entered in this fashion. As the abstracting progresses, items from different parts of the take-off will be amalgamated under a single common description. Related work section items from different parts of the take-off will begin to appear on the same abstract.

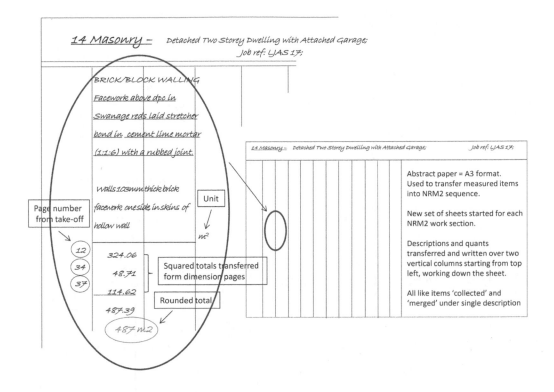

Figure 2.33 Layout, spacing and presentation of abstract.

The abstracted items should be spaced well apart allowing for the later insertion of any measured items that have been overlooked or omitted. Once all measured items have been transferred to the abstract, the quantities are totalled and rounded to the nearest whole unit (NRM2 3.3.2.e).

2.12 Billing

This is the final stage in the preparation of the completed BQ. The effort of assembling and ordering was completed when abstracting, and all that remains is for the descriptions and quantities to be presented in a structured and consistent fashion. Where NRM2 is used, the sequence of this presentation will follow the Work Sections 1 to 41 as detailed in NRM2 3.2.3.1. Within each Work Section the items that require measuring are listed (first column). This is followed (reading across the page from left to right) by the adopted unit of measurement (second column). This in turn is followed by three levels of heading for each item of work that requires measuring. These three levels are labelled as such in the Work Sections of NRM2. When structuring a BQ it would be possible to adopt this format in order to present the measured work component of a finished BQ. The following example shows this approach for NRM2 Work Section 5 (Figure 2.35 and page 40).

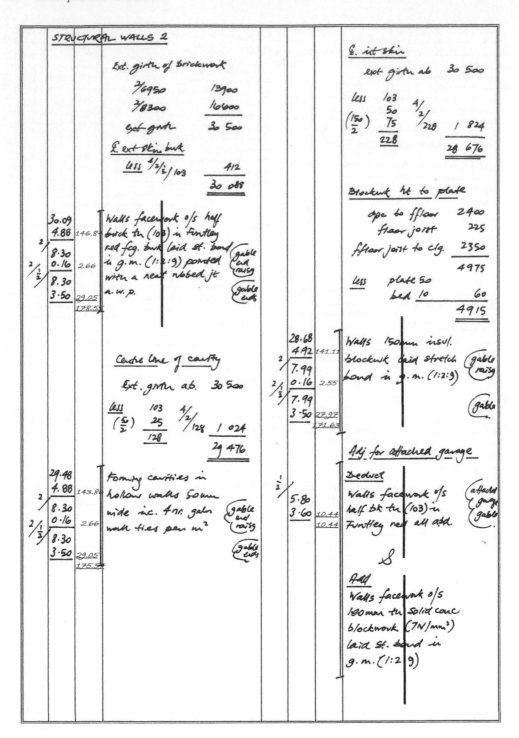

Figure 2.34 Completed page of dimensions showing line-through following transfer to abstract.

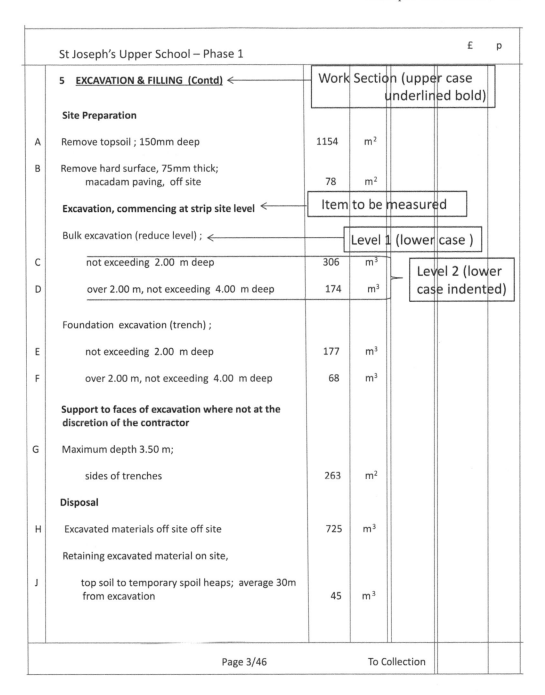

St Joseph's Upper School – Phase 1

£ p

	5 **EXCAVATION & FILLING** (Contd)				

Work Section (upper case underlined bold)

	Site Preparation		
A	Remove topsoil ; 150mm deep	1154	m²
B	Remove hard surface, 75mm thick; macadam paving, off site	78	m²

Excavation, commencing at strip site level

Item to be measured

Bulk excavation (reduce level) ;

Level 1 (lower case)

C	not exceeding 2.00 m deep	306	m³
D	over 2.00 m, not exceeding 4.00 m deep	174	m³

Level 2 (lower case indented)

Foundation excavation (trench) ;

E	not exceeding 2.00 m deep	177	m³
F	over 2.00 m, not exceeding 4.00 m deep	68	m³

Support to faces of excavation where not at the discretion of the contractor

G	Maximum depth 3.50 m;		
	sides of trenches	263	m²

Disposal

H	Excavated materials off site off site	725	m³
	Retaining excavated material on site,		
J	top soil to temporary spoil heaps; average 30m from excavation	45	m³

Page 3/46   To Collection

Figure 2.35 Presentation of completed BQ.

Excavating & filling.
Work Section: (e.g. 5 excavating & filling).
Item to be measured: (e.g. site preparation).
Level 1: (e.g. bulk excavation).
Level 2: (e.g. not exceeding 2.00m deep).
Level 3: (e.g. obstructions, manholes and the like that must remain undisturbed).

The use of upper and lower case letters together with underlining, bold face print and indenting helps to clarify this priority in the completed BQ. The preparation and presentation of a BQ is considered further in Chapter 4, Document production.

3 Mensuration

3.01 Introduction

Mensuration is the term used by mathematicians to describe the techniques used to establish lengths, areas and volumes. It is necessary to understand the principles of mensuration before dimensions can be correctly presented and recorded on dimension paper. While many people are unfamiliar with the term 'mensuration', most of the geometric formula is generally well known. This chapter provides an introduction to the more commonly adopted mensuration techniques practised by measurers. Examples of recording dimensions for geometric formula are given in Chapter 2.04 and Appendix II.

3.02 Girths

One of the most frequently used techniques when booking dimensions is 'girthing'. Most buildings are based on a square or rectangular plan shape and it is often necessary to establish the perimeter length of individual rooms or whole buildings either internally or externally. Drawings will show plan dimensions, but before these can be set down and recorded on dimension paper it will be necessary to 'build-up' perimeter lengths as a waste calculation. For example, if we wanted to measure the skirting or coving in a room, it would be convenient for the dimensions of the room to be given in the dimension column as a single linear item (i.e. the perimeter). This is carried out as a waste calculation, which clearly identifies the individual plan lengths as given on the drawing. It would be incorrect to carry out this calculation as mental arithmetic (in the head) and simply assume that everyone knew exactly

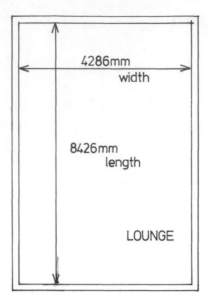

Figure 3.1 Plan of room showing internal room dimensions.

what you had done. One of the basic rules of recording dimensions is that these should be clearly presented so that they can be traced back to the drawing.

The waste calculation to establish the internal girth of the outline room plan above may be presented in one of two ways (Figures 3.2 and 3.3).

In the examples above (Figures 3.1, 3.2 and 3.3), the girth of the room was used to establish the length of the skirting. It might equally well have been used as the length dimension to establish the internal wall area (Figure 3.4).

In this example, the unit of measurement is square metres but the booked dimensions are based on the room's girth. If the girth is used in combination with the floor-to-ceiling height, this will provide the internal wall area. In this case it is applicable to two separate measurable items, wall plaster and emulsion paint, and both are linked to their common dimensions by an ampersand and bracket line.

If the external girth is required and the drawn details only show internal dimensions, adding the wall thickness to each plan dimension (once at each end) will give the external plan dimensions. Using the same girthing technique, the external perimeter length can be presented as a waste calculation (Figure 3.5).

3.03 Centre lines

When measuring brickwork or trench excavation it is necessary to book dimensions based on average or mean lengths. This is achieved by an adaptation of the girthing technique demonstrated earlier. This approach is known as a centre line calculation. To date we have measured areas that have been based on the girth or perimeter length of the room in question (such as plasterwork or decoration to walls). Other work sections of NRM2 require a length

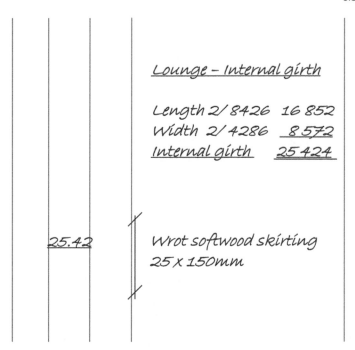

Figure 3.2 Recording 'room girths' as waste calculation (a) first presentation.

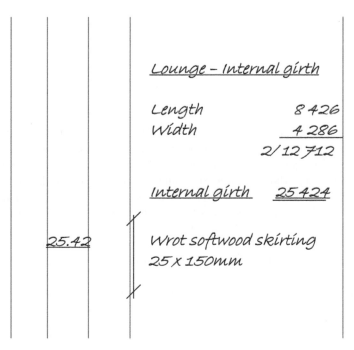

Figure 3.3 Recording room girths as waste calculations (b) alternative presentation. Both show the dimensions as they appear on the drawing (allowing anyone else to check them against the drawing). External girths are established in the same way.

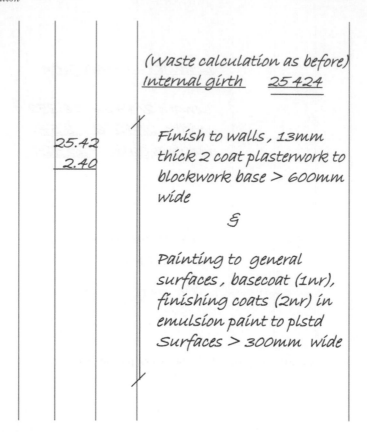

Figure 3.4 Establishing internal wall area based on girth calculation.

measurement based on the midpoint of the material being recorded. This is perhaps best explained by the example of a foundation trench excavation. Most foundation trenches for low-rise residential property are either 450 or 600 mm wide. In order to measure these accurately it is necessary to base all dimensions on the centre of the foundation trench. Another way of thinking of this is the average length of the foundation trench. In a situation where we are presented with internal plan dimensions, an internal girth can be found. If the internal plan dimensions are adjusted by adding half the total wall thickness at both ends, the centre line of the wall can be found. This last point is based on the reasonable assumption that the external walling is central to the width of the foundation. If it weren't, the building would be structurally unstable (Figures 3.5 and 3.6).

Rather than thinking of the technique in terms of adding to individual plan dimensions, it is more appropriate to consider the effect of this adjustment at each corner. This is only because it helps to understand the presentation of the waste calculation. If we consider the four corners of a regular plan-shaped building, the calculation involves 'timesing' by four, which in this case (since internal dimensions were used) means the result is added to the girth to give the centre line. Had an external girth been used, the same result would be deducted to provide the centre line (see Figure 3.6).

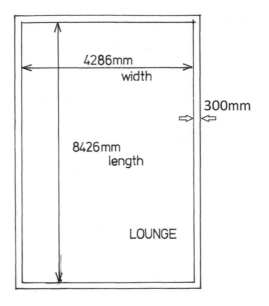

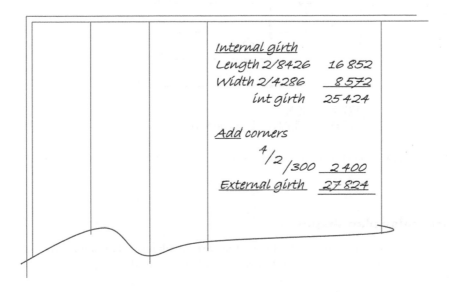

Figure 3.5 External girth (waste calculation) based on internal plan dimensions and wall thickness.

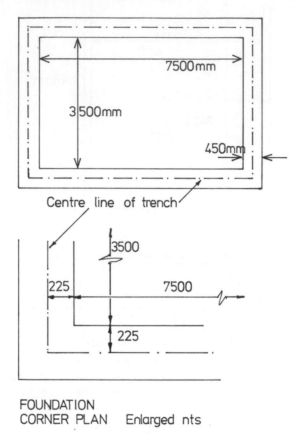

FOUNDATION PLAN   nts

7500mm

3 500mm

450mm

Centre line of trench

3500

225   7500

225

FOUNDATION
CORNER PLAN   Enlarged nts

Figure 3.6 Foundation trench plan and corner detail.

3.04 Irregular plan shapes

So far we have only considered buildings that are regular in plan shape. In many cases buildings are designed with clipped corners and insets, and these must be taken into consideration when establishing girths and centre lines (Figure 3.8).

In the case of a clipped corner, providing the angles remain at 90° there is no need for any adjustment in either the girth of the building or its centre line (Figure 3.9).

In Figure 3.9, the solid line shows the actual plan outline of the building and the hatched outline the line of a regular-shaped building. It will be seen that the corner could be 'folded out' to form a regular right-angled corner. In any girthing exercise the clipped corner has no effect, since an internal angle compensates for an external angle. This rule of compensation may be applied whatever the shape of a building, provided that the corners are all right angles. From this example it can be seen that, in regard to girth, an internal angle compensates for an external angle.

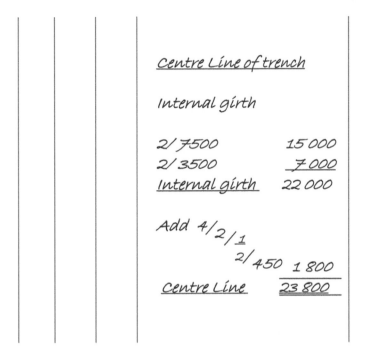

Figure 3.7 Waste calculation establishing length of trench (see Figure 3.6 foundation trench plan).

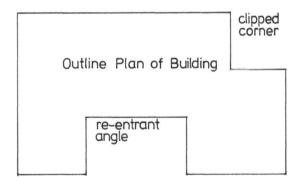

Figure 3.8 Clipped corners and re-entrant angles.

Figure 3.10 shows an irregular plan shape with external angles marked 'x' and internal angles marked 'o'. A simple tally of the external and internal angles will always give an excess of four external angles, providing the angles are all regular and the outline of the building is fully enclosed. The dimensions used to establish the girth of this building (Figure 3.9) need only relate to the overall lengths A and B. In regard to centre-line calculation, the equation need only make an adjustment for an excess of four external angles.

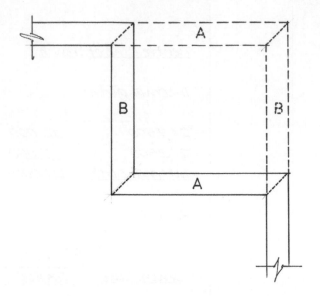

Figure 3.9 Clipped corner and girthing.

IRREGULAR OUTLINE PLAN

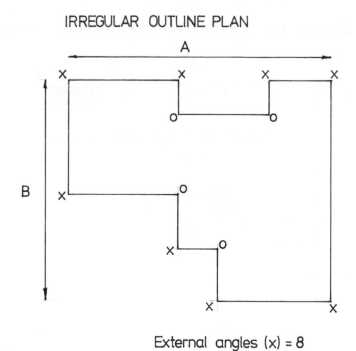

External angles (x) = 8
Internal angles (o) = 4
Excess of external angles = 4

Figure 3.10 Irregular outline plan.

3.05 Re-entrant angles

In this case (Figure 3.11) the plan shape is interrupted by a projection into the building (e.g. where a chimney stack projects into a room).

This will have an effect on the perimeter length of the room. The calculation of the girth will include the length L, above, but will make no allowance for the depth of the projection, D. These must be added to obtain the perimeter length. The waste calculation would be presented as in Figure 3.12.

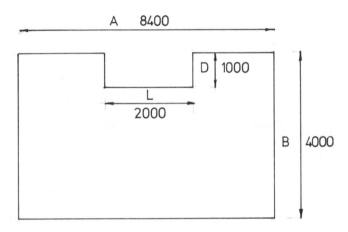

Figure 3.11 Dimensioned room plan showing re-entrant angle details.

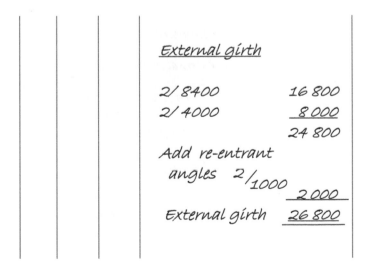

Figure 3.12 Waste calculation establishing perimeter length including re-entrant angle.

Care must be taken to study the drawings when calculating perimeter lengths. In the majority of cases the perimeter calculation will be used as the basis for recording a number of dimensions relating to it. Often the walls of an extension to a building or an external garden wall will not form a complete enclosure, and in these cases any adjustment for a centre line must take into account the number of external and internal angles (Figures 3.13 and 3.14).

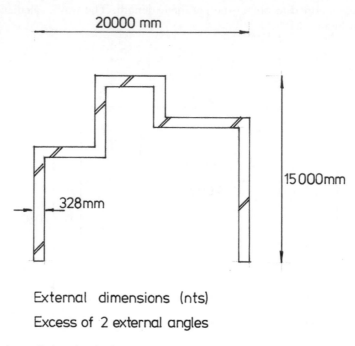

External dimensions (nts)

Excess of 2 external angles

Figure 3.13 Garden wall plan view (nts).

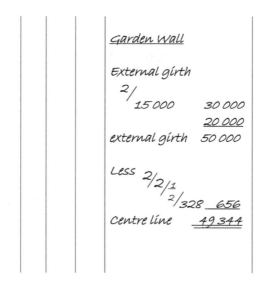

Figure 3.14 Centre-line calculation for garden wall.

In Figure 3.13 the walling shows four external angles and two internal angles. Once the girth is established, the adjustment for the centre line of the wall must allow for an excess of only two external angles (see Figure 3.14). An example of a measured boundary wall is included as part of the example take-off for External Works in Chapter 15.

3.06 Irregular areas

By carefully dividing large, irregular-shaped areas it is possible to establish a number of smaller geometric forms. These can then be recorded as a series of triangles, rectangles and squares (Figure 3.15).

This approach works well when the sides of an irregular-shaped area are straight. On occasion, a curved boundary line will require an adaptation of this approach (Figure 3.16).

IRREGULAR AREAS

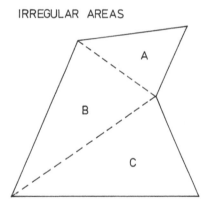

Figure 3.15 Irregular areas.

IRREGULAR AREAS

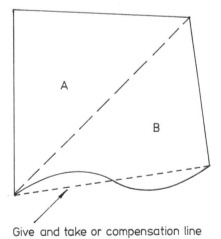

Give and take or compensation line

Figure 3.16 Compensation (give-and-take) line.

In order to approximate the area of this irregular area, the measurer will need to draw a straight line along the curved boundary to balance the areas inside and outside of the line (Figure 3.16). Having established the 'give-and-take line', the irregular area can be recorded using the same approach as before, as a series of triangles, rectangles and squares.

This is a rather haphazard approach which can be somewhat inaccurate. For a more accurate result and where evenly spaced offsets are available, the area may be divided into an even number of equally spaced strips and Simpson's rule applied.

The quickest and most accurate approach is to use a digitiser; this allows the measurer to enter the appropriate scale and simply run a digitiser pen around the outline of the irregular area. The result is almost simultaneously computed and presented on the screen.

3.07 Triangles and circles

Refer to Chapter 2 (2.04) 'Geometric forms' for examples of mensuration techniques associated with the measurement of triangles and circles.

3.08 Parallelograms and trapeziums

Where the two opposite sides of a four-sided shape are parallel (parallelogram), the area can be recorded by multiplying the perpendicular height (h) by the length (l) (Figure 3.17).

Where only two sides of a four-sided shape are parallel (trapezium), the perpendicular height between the parallel sides multiplied by the mean length of the parallel sides will give the area (Figure 3.18). The formula for each of these, together with a number of other geometric forms, is given in Appendix II.

3.09 Bellmouths

This term is used to describe the shape formed at a 'T' junction in a road or turning head in a cul-de-sac (Figure 3.19).

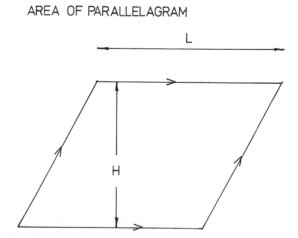

AREA OF PARALLELAGRAM

Figure 3.17 Area of parallelogram.

AREA OF TRAPEZIUM

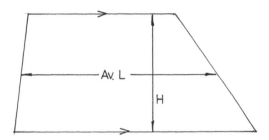

Figure 3.18 Area of trapezium.

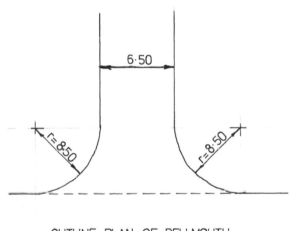

OUTLINE PLAN OF BELLMOUTH

Figure 3.19 Outline plan of bellmouth.

In the first instance the road surface will be measured through the bellmouth at 6.50 m width. This leaves two irregularly curved areas at each side of the 'T' junction. By initially recording two squares, each of 8.50 m side, and then deducting a quarter circle of 8.50 m radius (again on each side), the correct area will remain. In accordance with the basic principles of measurement, it is important to over-measure in the first instance and make adjustments after (Figure 3.20).

There are a number of other mensuration techniques that are used by measurers when taking-off quantities. Many of these are specific to certain types of work (e.g. groundworks, roof structures and roof coverings). Rather than describe these here, they are each included as part of the explanation of the measurement techniques appropriate to their respective chapters.

THREE STAGES FOR MEASUREMENT
OF A BELLMOUTH

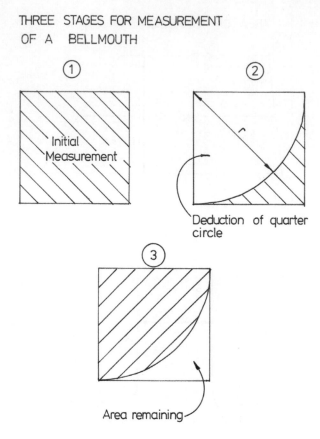

Figure 3.20 Three stages for measurement of a bellmouth.

4   Document production

4.01 Traditional BQ production

As we have already identified, the principal purpose of measurement is to make available a Bill of Quantities (BQ), which will be used as the basis for the preparation of tenders for construction work. In due course the priced BQ will become one of the contract documents that legally bind the contractor to perform and the client to pay for the construction. In this form, the BQ provides a basis for the valuation of varied works and the preparation of interim or stage payments. It is also likely to be used by the contractor in the administration and organisation of construction operations.

Measurement can therefore be described as the starting point from which construction costs are established. In order for measured items to accommodate the costing process, they must be framed in a way that is meaningful to the sequence and division of operations on site. Costing a document comprising sporadic and unstructured measured items would be impossible.

Tendering arrangements may vary but in the normal course of events the contracted works will be performed as a number of subcontract operations, and it is important that the costing documentation replicates this pattern and distribution of work. There are a few anomalies but, by and large, the divisions of subcontracting operations are represented in this fashion by the Work Sections of NRM2. Other documents may (or may not) be prepared in accordance with the same structure of Work Sections as given in NRM2. When this is the case, the respective documents can be cross-referenced to each other, avoiding unnecessary repetition and providing a coherent structure for all tender documents. Previous editions of the Standard Method of Measurement (SMM7) provided the opportunity to coordinate project information in this way, but there has been no attempt to do this with NRM2.

Prior to the wide-scale availability and adoption of computer technology, the preparation of the BQ, by necessity, involved a number of tedious and repetitive operations culminating in the preparation of a handwritten draft. Once typed, and before being copied, the master would be read over. The completed document would then be distributed with other tender documents to tendering contractors. There were a number of attempts made to rationalise this procedure. Billing direct and cut-and-shuffle are examples of manual systems that were developed in order to avoid the need for abstracting.

Technology is available today (5D BIM) which can produce schedules of work and associated construction costings in minutes rather than months (see Chapter 4.02). As such, this eliminates the need to record dimensions, observe the protocols of booking dimensions or follow any standard method of measurement. The reader may even question why a textbook such as this is necessary? Regardless of the efficiencies that these emerging technologies are able to offer (and they may be considerable), all of this presumes that the user of the software is capable of interpreting the input/output produced by the BIM package. Without basic principles being in place there is the risk that significant shortcomings may be overlooked.

4.02 Building Information Modelling (BIM)

The process of producing a building (construction) is regarded as a complex operation that requires the successful integration of a number of different responsibilities and services. In addition, many of the factors that determine a successful outcome are operationally interdependent. The combination of this uncertainty and complexity has historically been regarded as an industry-specific issue. The idea that a single system would be able to embody the multiple layers of complexity that are typical of any construction project had, until recently, largely been discounted.

4.02.01 The idea behind BIM

The progress of what we term 'Building Information Modelling' could only have been possible with the adoption of computer-aided design (CAD) and the digitalisation of hard copy drawn information and written specification. Once it became possible to represent design information digitally, the precedent for developing a single model for a construction project had been set.

It has taken some time for the idea of a building model to become established, but the potential efficiencies had been noted by the UK Government with a significant investment of public funds in the construction and civil engineering sectors. In order to facilitate progress towards greater efficiency, the UK Government identified distinct and recognisable milestones in the incremental implementation of BIM in the form of 'levels'. These levels were initially established within the range 0 to 3, where zero represented little recognition for the BIM process and three reflected full collaboration and a shared project model. The interpretation of these levels is not universally established. For the purposes of this text the following interpretation is offered.

4.02.02 Levels of BIM

4.02.02.01 Level 0 BIM

Level 0 effectively means no collaboration. This is typified by 2D CAD drafting mainly for Production Information (RIBA Plan of Work 2013 stage four). Output and distribution are via paper or electronic prints, or a mixture of both. The majority of the construction industry is in advance of this level (source: NBS National BIM Report 2014).

4.02.02.02 Level 1 BIM

This is characterised by a mixture of 3D CAD for concept work, and 2D when drafting building regulation, planning approval documentation and production information. CAD standards are benchmarked to BS 1192:2007, with electronic sharing of data coordinated from a common data environment (CDE). There is a distinct lack of collaboration between the different disciplines with each party retaining responsibility for maintaining and publishing their own data.

4.02.02.03 Level 2 BIM

This can be identified by collaboration between all parties. While each may use their own 3D CAD model, they may not necessarily be working on a single shared model. Instead, design information is shared among the different parties through a common file format. Initially this data set may have been established and developed by any member of the design team, including the client, quantity surveyor, architect, structural engineer, building services engineer, supplier, contractor or subcontractor. These separate data are collectively pooled in order to make a 'federated BIM model', and it is only at this point that it becomes possible to carry out interrogative checks. The whole process is dependent on the protocol of common file formats such as IFC (Industry Foundation Class) or COBie (Construction Operations Building Information Exchange). There is an expectation that, by 2016, this will be the minimum target for all UK government public sector work.

4.02.02.04 Level 3 BIM

Sometimes referred to as 'iBIM' or integrated BIM, this is the UK Government's target for public sector work beyond 2016. Level 3 BIM assumes full collaboration and a shared project model located in a centralised repository. All parties have access and are able to make amendments to the model. Clash detection allows the design team to ensure that, for example, structural components and air-conditioning ducting do not occupy the same space. The very fact that the model is 'live' (at least in principle) means that clash detection becomes unnecessary. This level of collaboration is enhanced by the 'interoperability' of the software: that is, the ability of any system (or software) to be understood by, and interface with, other systems or software.

Once the building model has been created, specialist software is able to provide a realistic image of the design. This process of rendering allows the client, the design team and end-users to experience a 3D internal and external view of the proposal. 3D walk-throughs of the finished building enable an interactive experience months before work on site has commenced. While the potential of level 3 BIM is generally recognised, it would be wrong to assume that there will be a natural progression towards the universal adoption of this technology as an industry standard.

The level of collaboration required to achieve level 3 BIM has raised issues among the various parties involved in construction projects. Particular concerns have been raised with regard to design ownership, liability and legal interpretation. Some progress has been made towards finding a resolution to these concerns (CIC BIM Protocol). Even so, at the time of writing, construction professionals and the construction industry generally are aware that there are areas that remain unresolved, such as copyright and liability.

4.02.03 Beyond level 3 BIM

Once a building has been modelled, the data thereby created has the potential to be extracted and interrogated in many different ways.

4.02.03.01 4D BIM (Time)

Four-dimensional (4D) BIM (Time) provides the opportunity to input construction planning and management to the BIM model and is able to generate a project delivery timeline, which in turn can enable the automated scheduling of resources and quantities. Tools can further be used to enhance the planning and monitoring of health and safety requirements as the project progresses. Should the design change, advanced BIM models will be able to automatically identify these changes and map their effect on the critical path, indicating a corresponding impact on the overall delivery of the project.

4.02.03.02 5D BIM (Cost)

Five-dimensional (5D) BIM (Cost) may be the 'holy grail' so far as the QS profession is concerned. This can be described as the capacity to integrate design information with the ability to record dimensions (taking-off) together with the facility to quantify and cost the project. Changes to the design will automatically adjust quantities and costs.

4.02.03.03 6D BIM (Post-delivery facilities management)

Six-dimensional (6D) BIM is the building model as it is delivered to the construction owner upon completion of the project. It provides the client with all of the necessary information/documentation to ensure that the building performs as it was intended and is maintained in order to maximise efficiency throughout its whole life. This includes operation, end-of-use demolition and recycling. As part of the suite of NRM documents, the New Rules of Measurement 3 (NRM3) are intended to provide a consistent method for the quantification and description of maintenance works.

While the industry aspires towards level 3 BIM, the reality remains that the majority of built environment professionals have yet to adopt a BIM-based approach. Even so, a significant minority of UK QS practices and contractor organisations have made progress towards automating the traditional (and time-consuming) process of producing a BQ. Regardless of developments in technology and software, it is important that all involved in the preparation of construction project documentation have a thorough grasp of the structure, coding and presentation of a BQ. The last part of this chapter explains the techniques and processes associated with traditional (non-BIM-compliant) BQ production.

4.03 The structure and presentation of the BQ

Bills of Quantities should 'fully describe and accurately represent the quantity and quality of the works to be carried out'. Providing they are able to achieve this, there would appear to be no limitation on the eventual form. While recognising the need to adopt a flexible approach, the end product must provide a document with a recognisable structure that can be consistently interpreted by all users. A typical BQ would comprise a number of separate sections including Preliminaries, Preambles, Measured Work, Work by Named Subcontractors/Suppliers and Non-measurable works (Provisional Sums). NRM2 identifies these last two items as part of the Preliminaries, but it is often found convenient (for example when preparing valuations and final accounts) if these are grouped together in a separate section of the finished BQ.

4.03.01 Bill preparation

In order to achieve consistent interpretation by all who use the BQ, a standard approach to billing will be necessary. While the format and style of presentation may vary from one office to the next, the general principles remain the same. Standard Ruled A4-size paper (known as single bill paper) is used for the presentation of billed work (see Figure 4.03). The common availability of word processing software has enabled the typical QS practice to develop variants to the standard BQ page. While this has facilitated a more flexible interpretation in both style and format, it should not detract from a consistent interpretation by all who use the BQ (Figure 4.1).

The finished BQ should be presented with a cover that includes the title of the project, the name of the client, details of the QS practice responsible for the project and the date (Figure 4.2).

Historically the finished tender documents, including the BQ, have been presented as hard copy. Where legal requirements permit, there would seem no reason why these cannot be made accessible electronically or online. The submission of e-tenders and particularly the situation regarding offer and acceptance is likely to require legal guidance.

Care should be taken to ensure that all relevant documents, including the BQ, drawings, schedules, specification and all Mandatory Information (as required by NRM2), accompany the tender documents. It is essential that all tenderers receive an identical set of tender documents.

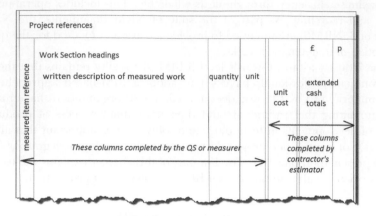

Figure 4.1 Standard rulings for Bill of Quantities.

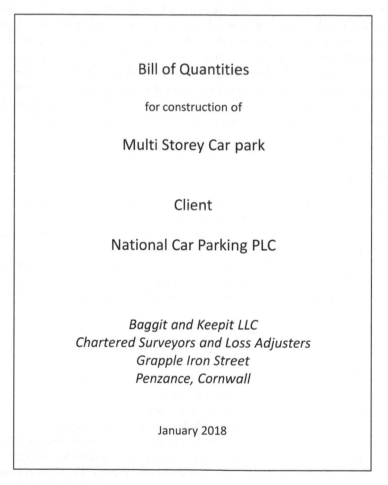

Figure 4.2 Example Bill of Quantities front cover.

St Joseph's Upper School – Phase 1					£	p
	5 EXCAVATION & FILLING (Contd)					
	Site Preparation					
A	Remove topsoil ; 150mm deep	1154	m²			
B	Remove hard surface, 75mm thick; macadam paving, off site	78	m²			
	Excavation, commencing at strip site level					
	Bulk excavation (reduce level) ;					
C	not exceeding 2.00 m deep	306	m³			
D	over 2.00 m, not exceeding 4.00 m deep	174	m³			
	Foundation excavation (trench) ;					
E	not exceeding 2.00 m deep	177	m³			
F	over 2.00 m, not exceeding 4.00 m deep	68	m³			
	Support to faces of excavation where not at the discretion of the contractor					
	Maximum depth 3.50 m;					
G	sides of trenches	263	m²			
	Disposal					
H	Excavated materials off site	725	m³			
	Retaining excavated material on site,					
J	top soil to temporary spoil heaps; average 30m from excavation	45	m³			
	Page 3/46			To Collection		

Figure 4.3 Completed BQ page.

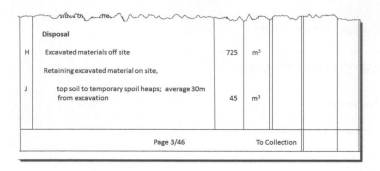

Figure 4.4 Foot of BQ.

Each Work Section would be commenced on a new sheet. The extreme left-hand column (binding margin) is used to provide a system of indexing. This is necessary to allow all those who use the bill to locate and refer to specific bill items, and is normally achieved by entering consecutive letters of the alphabet parallel with the first line of the description (avoiding o and i). When used in combination with the BQ page number, this provides a unique reference for each bill item. Each page of the BQ should give the job name and reference at the head of the page while the cash columns at the top on the extreme right should be labelled with the appropriate Work Section and the cash columns titled with (£, p) pounds and pence (Figure 4.3).

The start of each BQ should give the bill number and the appropriate Work Section heading. The use of upper and lower case letters, bold face, underlining and indenting can provide a structure and sequence to the descriptions (Figure 2.35). Assuming the adoption of a traditional BQ, the sequence of bill items will follow NRM2 Work Section order. Descriptions should be written in full to avoid the possibility of misinterpretation.

4.03.02 Units of billing

As part of the abstracting process, the quantities will have been rounded to the nearest whole unit. Fractions of less than a half of a unit will have been disregarded, while those over a half will be taken as a whole unit. In some circumstances this may result in the elimination of a measured item (i.e. where the unit is less than 0.50 of a whole unit), in which case the item will be given as a whole unit (NRM2 3.3.2.1.e). Exceptionally, quantities measured in tonnes should be given to two places of decimals. When entered in the unit column the following symbols are used to represent these units:

ha   = hectare
hr   = hour
kg   = kilogram
kN   = kilonewton
kW   = kilowatt
m   = linear metre
m^2 = square metre

m^3 = cubic metre
mm   = millimetre
mm^2 = square millimetre
mm^3 = cubic millimetre
nr   = number
t   = tonne
wk   = week.

At the foot of each page the cash columns should be ruled off and the words 'To Collection' entered. At the end of each work section a Collection is set up to provide a framework for the estimator to total the costs for each work section (Figure 4.4).

At the end of the BQ, a summary of the totals from all of the work sections is prepared. The total cost of the measured work will in turn be transferred to a General or Grand Summary, which collects the costs from the other bill sections and provides the total to be transferred to the form of tender (Figure 4.6).

The component parts of the BQ may include all or some of the following:

Preliminaries
Preambles
Measured Work
Provisional sums
Defined
Undefined
Risks (contingencies).

These may be presented as separate sections or, in some cases, combined to provide conglomerate sections. This will depend on personal preference or the particular requirements of the project. Where the structure of NRM2 work sections is adopted, this can be set up to include Provisional Sums, both defined and undefined and work by Named Subcontractors, Suppliers and Statutory Undertakers with the relevant NRM2 work sections. Alternatively these may each be given separately. The cost allocation of any identifiable risk (as defined by NRM2 2.10) may also merit a discrete BQ section. Each part of the BQ would be represented by a section number (e.g. Section 1 Preliminaries, Section 2 Preambles, Section 3 Measured Work Sections, etc.). Page numbers relate to each Section, and thus if there were 24 pages of Preliminaries these would be identified at the foot of each page by the Section number followed by the page number (1/1 to 1/24). Each description is identified by a reference, either alphabetic or numeric, which when used in combination with the page number provides a unique location reference. All Work Sections are followed by a Collection or Summary page, which facilitates the transfer and totalling of the previous page totals (Figure 4.5). Each of the Measured Work Sections requires an individual Work Section Summary (Figure 4.6) that will, in turn, be transferred to a General Collection Page (Figure 4.7).

4.03.03 Preliminaries (NRM2 Work Section 1)

The introduction of NRM2 has allowed the opportunity to update and enhance the detail of costing associated with Preliminaries. The function and purpose of Preliminary cost remain unchanged in that they allow for the costing of items that are not part of the permanent works

Samuel Thomas Design Studio – Machynlleth, Powys	£	p

14 MASONRY

<u>COLLECTION</u>

Page	3/61
Page	3/62
Page	3/63
Page	3/64
Page	3/65
Page	3/66
Page	3/67
Page	3/68
Page	3/69
Page	3/70
Page	3/71

TO SUMMARY   £

Page 3/72

Figure 4.5 BQ work section collection page.

Sam Thomas Design Studio – Machynlleth, Powys		£	p
SUMMARY	Page		
3 DEMOLITION	3/ 7		
5 EXCAVATION & FILLING	3/15		
6 GROUND REMEDIATION AND SOIL STABILISATION	3/26		
11 IN-SITU CONCRETE WORKS	3/30		
14 MASONRY	3/38		
16 CARPENTRY	3/49		
18 TILE AND SLATE ROOF AND WALL COVERING	3/55		
22 GENERAL JOINERY	3/62		
23 WINDOWS, SCREENS AND LIGHTS	3/70		
24 DOORS, SHUTTERS AND HATCHES	3/76		
25 STAIRS, WALKWAYS AND BALUSTRADES	3/81		
27 GLAZING	3/84		
28 FLOOR, WALL, CEILING AND ROOF FINISHINGS	3/88		
29 DECORATION	3/96		
31 INSULATION, FIRE STOPPING AND FIRE PROTECTION	3/104		
33 DRAINAGE ABOVE GROUND	3/107		
34 DRAINAGE BELOW GROUND	3/112		
35 SITE WORKS	3/120		
TOTAL CARRIED TO GENERAL SUMMARY		£	
Page 3/121			

Figure 4.6 BQ Measured work summary page.

Project title/ref					£	p
General Summary	Page					
SECTION 1 – PRELIMINARIES	1/24					
SECTION 2 – PREAMBLES	2/48					
SECTION 3 - MEASURED WORK	3/156					
SECTION4 - NAMED SUB CONTRACTORS	4/14					
SECTION 5 - CONTINGENCY/RISK	5/3					
TOTAL CARRIED TO FORM OF TENDER			£			
Page 3/45			To Collection			

Figure 4.7 General or grand summary page.

but are nonetheless necessary for the successful completion of the project. Items such as insurances, temporary buildings, scaffolding, site hoardings, protection and the provision and maintenance of plant are all examples of such Preliminary items.

Assuming a main works contract, preliminaries are divided into two categories, information and requirements and the pricing schedule. The first of these (Part A) provides the project particulars and identifies the works, the project, the type of contract and the parties to that contract. The pricing schedule (Part B) would include provision for costing all of the items in the previous paragraph that are 'necessary for project completion but not a permanent part of the finished project'. The Preliminary Work Section of NRM2 has recognised that not all projects are let on a main contract basis, but instead may be project managed as a series of works packages or subcontracts. So a duplicate set of Preliminary clauses has been provided in order to accommodate different contractual arrangements. See NRM2 Preliminaries (main contract) and NRM2 Preliminaries (works package contract).

4.03.04 Preambles

The Preambles define the quality of materials and the standard of workmanship for the project. They may be included in the BQ as a separate conglomerate section or broken down and presented with the Work Section to which they relate. While they will inevitably affect the price of the measured work section, it makes for better presentation if they are included in the finished document as a discrete and separate BQ section. Where the architect/designer has prepared drawing specification references using the National Builders Specification (NBS), the measured work items and drawing references will automatically cross-reference to the Preambles. At the time of writing, NBS is developing a 'national BIM library' that has the potential to link specification clause to BIM model objects. NBS provides a valuable source of information in the preparation of both Specification and Preambles.

4.03.05 Measured Work Section

In most cases this will be the largest section of the BQ. For a BQ presented in Work Section order this will follow the sequence of NRM2 (see Figure 4.6). It is most unlikely that the completed BQ will include each and every NRM2 Work Section. Each project will generate an individual set of Work Sections that are specific to their individual circumstance. Examples of layout, page format and presentation of Measured Work Sections are included elsewhere in this chapter.

4.03.06 Prime cost sums

These include work executed by specialist subcontractors or material suppliers 'nominated' or selected by the architect or client. The contractor is entitled to recover any costs associated with the nomination, such as attendances and loss of profit, and the BQ is prepared to enable tenderers to make an inclusion for these. In order to complete the work, nominated subcontractors may require the use of temporary lighting, storage facilities and scaffolding. These are termed 'Contractor's General Cost Items' or 'General Attendance Items' and will be provided as a matter of course by the main contractor for the satisfactory execution of the works. Since these facilities are available anyway, the additional cost associated with nominated subcontractor use is minimal. However, they are technically the property of the contractor

Named Sub-Contractors				£	p
A	Include the prime cost sum of £4,800.00 for electrical power and lighting installation all in accordance with Electrical Engineer's specification C3298 and Drawing ref P3237			4,800	00
B	Allow for attendance	Item			
C	Add main contractor's profit	%			
	Page 3/45		To Collection		

Figure 4.8 Named subcontractor.

Work by Statutory Undertaker				£	p
	Include the following costs for work to be carried out by Statutory Undertakers. The Contractor is invited to add for profit and any loss of discount. The amount included in the final settlement will be based upon the actual charge made by the Statutory Undertaker.				
A	Eastern & Western Power			1,800	00
B	Terry Telecoms (new service line installation)			2,500	00
C	River Test is Best Water Co			3,000	00
	Page 3/45	To Collection			

Figure 4.9 Statutory Undertaker.

and the opportunity for these to be costed must be made at the tender stage. Where the nomination is for work of a specialist nature, or where it involves the provision of facilities which are not defined as general cost items, the BQ should make provision for the inclusion of 'Special Attendance items' that are additional to the main contractor's provision. It is possible for Prime Cost and Provisional Sums to appear in any one of three different locations in the completed BQ. They may be included with the Preliminaries, with the Measured Work Sections or in a separate section of their own. Wherever they appear, they would typically be presented as shown in Figures 4.8 and 4.9.

4.03.07 Provisional sums

Where there is insufficient information available for measurement, an undefined Provisional Sum can be written into the BQ. An inclusion may also be made where there is an element of doubt over actual costs but where the extent and nature of the work will become apparent once work on site has commenced. Two categories of Provisional Work are identified in NRM2 – defined and undefined. Assuming that there are sufficient data to define the timescale and to provide a construction programme, defined Provisional Sums can be allocated for the performance of specific tasks. The nature, method and extent of the work would only become apparent once work on site had commenced. Undefined Provisional Sums are more difficult to describe and would generally embrace sums of money written into the BQ for unforeseen circumstances by way of a contingency. This might include an allowance for the cost of any unforeseen additional substructures, the provision of samples for inspection or the carrying out of tests (Figure 4.10).

The next ten chapters offer an interpretation and explanation of booking dimensions in accordance with NRM2 for both low-rise residential construction and (in the case of structural concrete and steel frames) multi-storey buildings. With one or two exceptions, these are presented in the sequence of construction operations as they would occur on site. For any one new to the idea of recording dimensions, it is suggested they start with Chapters 2 and 3 followed by Chapter 12 (Floor, Wall and Ceiling Finishings).

Project details : Jemma P & Dani H Associates /ref xyz				£	p
<u>Defined Provisional Sums</u>					
Include the following defined provisional sums					
Supply only (fixing measured elsewhere)					
A	Kitchen wall and floor units			3,800	00
B	Sanitary Appliances			1,250	00
C	Air handling and heat recovery system			2,500	00
	Page 3/45	To Collection			

Figure 4.10 Defined provisional sums.

5   Substructures

5.01 Introduction

'Substructures' is the generic term used to describe construction operations below ground level, also referred to as foundations. In the normal course of events, the construction of foundations will require the excavation of trenches, the casting of concrete trench fill or strip foundations and the construction of brickwork or blockwork to support external walls and a ground floor slab. In addition to Excavation and Filling Section (NRM2 Work Section 5), the measurer will need to be familiar with the measurement rules for In situ

Concrete (NRM2 Work Section 11), Precast Concrete (NRM2 Work Section 13) and Masonry (NRM2 Work Section 14). The distinction between substructure and superstructure should be taken as the damp proof course and/or the damp proof membrane.

Within this broad definition of substructures, there are a number of other operations that might reasonably be described as providing structural support for the main body or superstructure of a building. These include basements, piling, underpinning, diaphragm walling, crib walls and gabions. There are also other NRM2 Work Sections associated with substructure operations, including Work Section 6 Ground Remediation and soil stabilisation. Consideration is given here for the substructure operations associated with domestic construction, including trench fill, strip foundations, reduce level excavation. The examples at the end of this chapter include trench fill foundations and basements.

5.02 Excavation

When measuring excavation for low-rise residential dwellings, this would typically involve booking dimensions for the following items: excavating and disposing of topsoil (two separate items), reduce level excavation + disposal, and excavating foundation trenches + disposal (see Table 5.1). In addition, it may be necessary to measure any of the following: temporary support to the sides of trenches, blinding concrete to the base of trenches, structural floors and the disposal of any groundwater that may be present during these operations.

The cost of excavation is dependent on a number of factors. Before commencing measurement it will be necessary to study the site survey drawings, the specification and any other information included on the site plans, together with borehole log details in order to identify the following.

- existing ground levels;
- finished ground levels;
- finished floor levels;
- foundation starting levels, depths and widths;
- groundwater level and the date of reading;
- evidence of any contamination or hazardous materials;
- restrictions/limitations on the disposal of surface and/or groundwater;
- the existence of known underground services or obstructions;
- details of any live overhead services crossing the site;
- the presence and levels of any substrata rock.

Table 5.1 Example for inclusion of Mandatory Information when booking dimensions.

Operation	Items to be measured	Unit of measurement	NRM2 clause
Excavate topsoil	1. Excavation 2. Disposal	m² m³ m³	5.5.2 5.9.2.3 5.10.1/2.1
Bulk excavation	1. Excavation 2. Disposal	m³ m³	5.6.1.1-3 5.9.2.1-3
Foundation excavation	1. Excavation 2. Disposal	m³ m³	5.6.2.1-3 5.9.2.1-3

NRM2.5 requires all of these data to be detailed as a mandatory inclusion in the completed BQ (NRM2.5 Mandatory Information to be provided). Initially this can be included as a heading in the take-off, which will subsequently be transferred to provide a heading in the appropriate Work Section of the BQ (Figure 5.1).

5.02.01 Surface (bulk) excavation

The term 'surface excavation' refers to the excavation of the surface of the site and describes the excavation of topsoil and/or bulk excavation to reduce levels. The surface of most 'green field' sites comprises a compressible layer of vegetable matter called 'topsoil'. Where space permits, topsoil can be carefully removed and stored on site for later reuse. This should be measured in square metres, stating the average topsoil depth in the description (NRM2.5.5.2). Alternatively, if there is no topsoil or if it is not required on site, it can be grouped with the bulk/reduce level excavation and measured in cubic metres, stating the maximum depth of dig in the description (NRM2.5.6.1.1–3).

The plan dimensions for any surface excavation associated with foundations must include the spread of the foundations beyond the face of the external walls. These may not be immediately obvious from the site plans, since the plan dimension on the drawings shows only external wall dimensions. (As far as measurement is concerned, it is only necessary to measure the minimum excavation required. In practice it is unlikely that a contractor will excavate within this minimum.) Working on the reasonable assumption that the walling is central to the

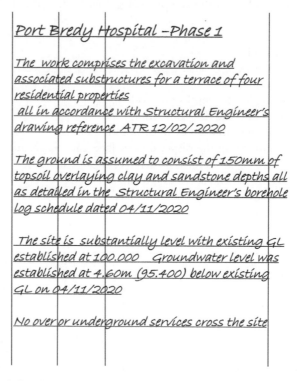

Figure 5.1 Port Bredy hospital – phase 1.

foundation, then the projection on either side of the foundation will be the same. Given the width of external wall foundations, together with the thickness of the external walling, it is therefore possible to calculate the projection of the foundation beyond the external wall. This would be presented on dimension paper as a waste calculation (see Figure 5.2).

Few sites are level, and the majority will require some excavation to provide a horizontal surface from which construction operations can commence. The removal of material in this way is measured in cubic metres by booking the plan dimensions together with the average

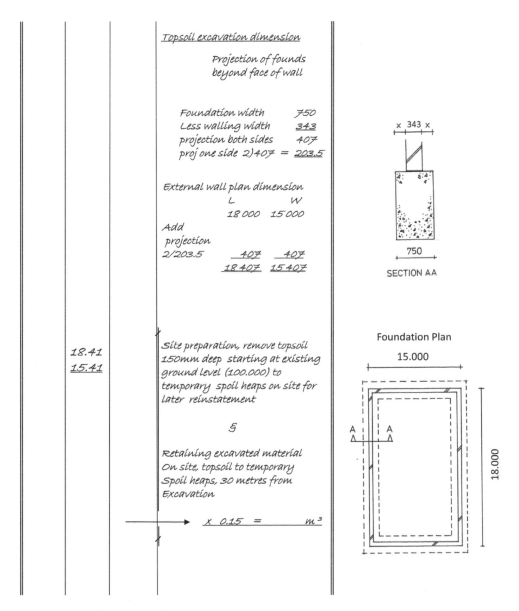

Figure 5.2 Topsoil excavation dimension.

depth of the bulk material that has to be removed. The plan dimensions for any reduce level bulk excavation should include the projection of foundations beyond the face of external walls. The depth will obviously vary and will need to be recorded in the dimension column as an average. Where a grid of levels is available, this can be carried out as shown in Figure 5.3.

A level is recorded at each intersection and each corner of the grid. If each individual grid square is averaged and the total of these subsequently averaged, the overall average ground level can be found. This can be an extended and cumbersome waste calculation. The same result can be found more efficiently, however, using the weighting method. Depending on

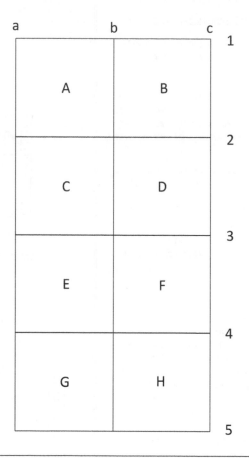

Depending on its position in the grid; each point will occur one, two or four times in the total averaging exercise. For example grid point 1b above occurs twice, once in the calculation of grid square A, and once in grid square B. Grid point 4b will is used four times, once for each grid square E, F, G and H. The grid point at the four corners will only be required once. Each grid level is listed in turn and then multiplied by the number of times it occurs in the overall calculation (i.e. its weighting). If this total is divided by the number of grid points used in the calculation, the result is the average existing ground level.

Figure 5.3 Plan view of the surface of a site with the overlay of a grid of levels.

its position in the grid, each grid point will occur one, two or four times in an overall ground-level averaging exercise. For example (see Figure 5.3), grid point 1b occurs twice, once in the calculation of grid square A and once in grid square B. Grid point 4b will be included four times, once for each grid square E, F, G and H. The grid point at the four corners will only be required once. Each grid level is listed in turn and then multiplied by the number of times it occurs in the overall calculation (i.e. its weighting). If this total is divided by the number of grid points used in the overall calculation, the result is the average existing ground level.

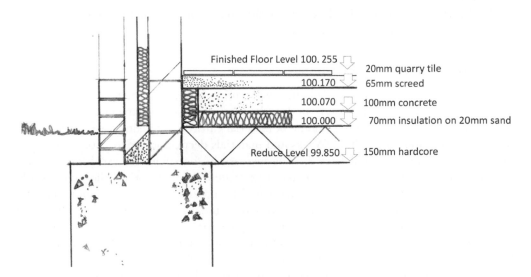

Figure 5.4a Section through insulated concrete ground floor showing relative ground levels.

Finished Floor Level 100. 255 — 20mm quarry tile
100.170 — 65mm screed
100.070 — 100mm concrete
100.000 — 70mm insulation on 20mm sand
Reduce Level 99.850 — 150mm hardcore

Ground floor Construction

Quarry Tile	20
Screed	65
Concrete bed	100
Insulation	70
Sand blinding	25
Hardcore bed	150
Ground Floor o'all th =	430

Finished Floor level   100.255

Less ground floor th   0.430
Reduced level =   99.825

Figure 5.4b Waste calculation to establish reduce level.

From a section through the proposed ground floor slab construction, the total slab construction thickness can be found. Once this is known, it is deducted from the finished floor level to provide the reduce level required. Having established the existing ground level and the required level, the average reduce level excavation depth can be found by deducting one from the other. Figures 5.4a and 5.4b show the waste calculations and subsequent booked dimensions for the removal of topsoil and bulk (reduce level) dig. The extreme dimensions of this over-site excavation are assumed to be 30 × 30 m and the reduce level required is given as 18.000 (Figure 5.5).

The description for reduce level excavation should give the maximum depth range, rather than the average depth required over the entire site.

5.02.02 *Disposal of excavated materials*

When any excavation work is carried out it will tend to increase its 'bulk', since air pockets are created in what was previously undisturbed ground. No adjustment is made to the quantities for this increase in the volume of material caused by 'bulking' (see NRM2 Work Section 5 Excavation and Filling 'Works and materials deemed included'; notes comments and glossary column General Rule 3). The increase in the volume of excavated material will vary, depending on the type of material excavated. An appropriate allowance will be made by the estimator, who will accommodate this apparent increase in volume once the substructure work is costed. Removing excavated material from site often costs more than the initial excavation. For this reason it is always necessary to measure disposal of excavated materials as a *separate item* (i.e. own description and quantities). NRM2.5.9.2.1–3 and NRM2.5.10.1–2.1–2 provide the alternatives for disposal of excavated materials.

When the BQ is prepared in accordance with NRM2, all excavation items will be classified as either bulk excavation or foundation excavation. The intention is to distinguish between excavation that can be carried out by front shovel blade (reduce level, basements, ponds, pools) and that by back-actor bucket (foundation trenches, pad foundations and pile caps). It is recommended that the specific detail of excavation type is included in the descriptive part of measurement, and this approach has been adopted in the preparation of this book. Each of these different types of excavation will in turn generate its own disposal item, which will be disposed of either on site (NRM2.5.10.1–2) or off site (NRM2.5.9.2.1–3). No reference is made as to where this material came from – trench excavation, basement excavation or reduce level excavation – it is all simply described as 'excavated material for disposal'. Where material is being disposed off site, it is necessary to identify the destination of this excavated material, together with any specific location and whether the material is in any way hazardous. (NRM2.5.9.2.1–3).

5.02.03 *Cut and fill*

As an alternative to excavating the surface of a site to obtain a level surface, a combination of excavating and filling can achieve the same end with the advantage of some savings in the cost of both excavation and disposal. This of course presumes that the excavated material is suitable for the purposes of filling and that adequate precautions are taken to ensure that the material is thoroughly compacted. Often the material will be unsuitable as fill, and any savings will be offset by the provision of imported fill material.

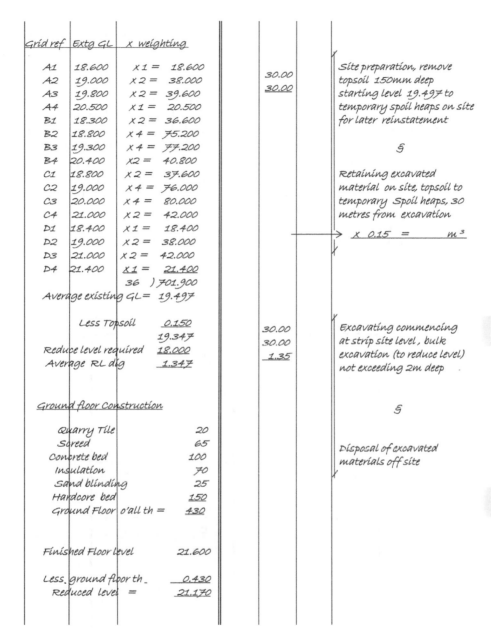

Grid ref	Extg GL	x weighting		
A1	18.600	x1 =	18.600	
A2	19.000	x2 =	38.000	
A3	19.800	x2 =	39.600	
A4	20.500	x1 =	20.500	
B1	18.300	x2 =	36.600	
B2	18.800	x4 =	75.200	
B3	19.300	x4 =	77.200	
B4	20.400	x2 =	40.800	
C1	18.800	x2 =	37.600	
C2	19.000	x4 =	76.000	
C3	20.000	x4 =	80.000	
C4	21.000	x2 =	42.000	
D1	18.400	x1 =	18.400	
D2	19.000	x2 =	38.000	
D3	21.000	x2 =	42.000	
D4	21.400	x1 =	21.400	

36) 701.900
Average existing GL = 19.497

Less Topsoil   0.150
                    19.347
Reduce level required   18.000
Average RL dig   1.347

Ground floor Construction

Quarry Tile   20
Screed   65
Concrete bed   100
Insulation   70
Sand blinding   25
Hardcore bed   150
Ground Floor o'all th =   430

Finished Floor level   21.600

Less ground floor th .   0.430
Reduced level =   21.170

30.00
30.00

Site preparation, remove topsoil 150mm deep starting level 19.497 to temporary spoil heaps on site for later reinstatement

&

Retaining excavated material on site, topsoil to temporary Spoil heaps, 30 metres from excavation

x 0.15 =   m³

30.00
30.00
1.35

Excavating commencing at strip site level, bulk excavation (to reduce level) not exceeding 2m deep

&

Disposal of excavated materials off site

Figure 5.5 Example of waste calculation and booked dimensions for a topsoil and reduce level excavation. Note to follow the above example: this exercise includes a waste calculation that would normally precede the booked dimension and is presented in this fashion simply because of the limitations of space.

When a site is to be cut and filled, the measurer should plot on the grid of levels a 'cut-and-fill' line. This line represents the point in a sloping site where there is no excavation or filling. On one side of the line the site is excavated, on the other it is filled. Where the excavated material can be used as over-site filling, the cost of disposal can be minimised (Figures 5.6 and 5.7).

In this example, a cut-and-fill line of 6.500 can be readily identified and positioned between the given grid points. Consider grid line 1: if the slope of the ground is assumed to be consistent then the contour line for 6.500 will fall half-way between the grid points 7.000 and 6.000. The same exercise is carried out on grid lines b and 2, where the cut-and-fill line coincides with the grid point at 3a. These points are then joined to give a contour line representing the cut-and-fill line. The site levels that are higher than the cut and fill line will be excavated, while those below will be filled. Each will need a waste calculation to determine the average dig depth and the average fill depth. Once these are known, the geometric forms representing the plan shape of the cut and fill can be recorded in the dimension column, followed by their respective average depths, thereby giving the volume of material to be either excavated or filled.

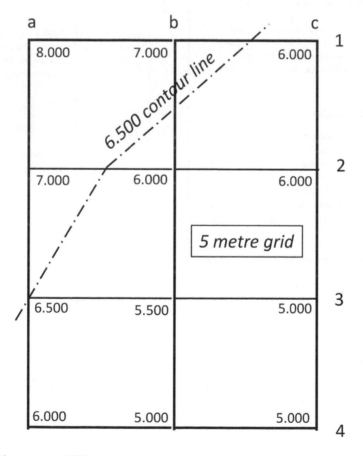

Figure 5.6 Plotting a cut-and-fill line.

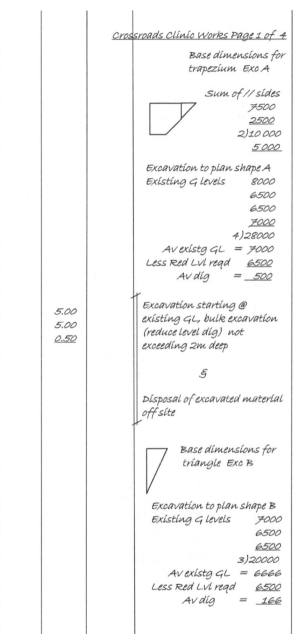

Base dimensions for
trapezium Exc A

Sum of // sides
7500
2500
2)10 000
5 000

Excavation to plan shape A
Existing G levels   8000
6500
6500
7000
4)28000
Av existg GL = 7000
Less Red Lvl reqd   6500
Av dig   = 500

5.00
5.00
0.50

Excavation starting @
existing GL, bulk excavation
(reduce level dig) not
exceeding 2m deep

&

Disposal of excavated material
off site

Base dimensions for
triangle Exc B

Excavation to plan shape B
Existing G levels   7000
6500
6500
3)20000
Av existg GL = 6666
Less Red Lvl reqd   6500
Av dig   = 166

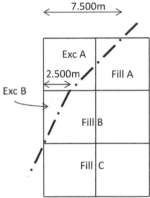

- 5metre grid
- Assume red level required = 6500

Five geometric forms can be identified
from the site plan reference Figure
5.06 as shown above

The plan dimensions for each of these
forms can be established by referring
to the contour line plot.

Each should be booked in turn having
first established the average
excavation or fill necessary

Figure 5.7 Crossroads clinic Works.

<u>*Crossroads Clinic Works Page 2 of 4*</u>

$\frac{1}{2}$ / 5.00
2.50
<u>0.17</u>

Excavation starting @ existing GL, bulk excavation (reduce level dig) not exceeding 2m deep

&

Disposal of excavated material off site

Base dimensions for Fill to C

Sum of // sides
   2500
   <u>7500</u>
  2)10 000
  = <u>5 000</u>

Filling to plan shape A
Existing G levels 6500
       6000
       6000
       <u>6500</u>
      4)25000
Av existg GL = 6250
Less Red Lvl reqd <u>6500</u>
  Av fill = <u>(250)</u>

5.00
5.00
<u>0.25</u>

Imported filling, beds over 50mm th but not exceeding 500mm deep, average 250mm finished thickness, level and to falls, cross falls or cambers

Figure 5.7 continued.

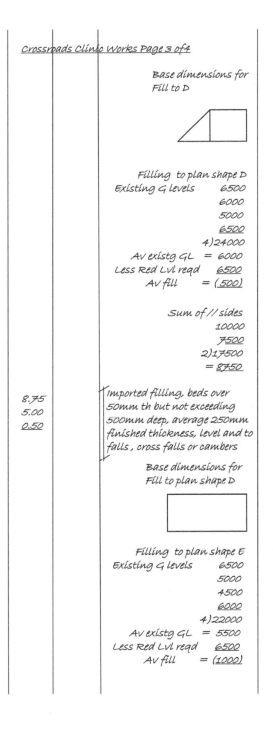

Crossroads Clinic Works Page 3 of 4

Base dimensions for
Fill to D

Filling to plan shape D
Existing G levels   6500
     6000
     5000
     6500
    4)24000
Av existg GL = 6000
Less Red Lvl reqd 6500
Av fill = (500)

Sum of // sides
   10000
   7500
  2)17500
  = 8750

8.75
5.00
0.50

Imported filling, beds over 50mm th but not exceeding 500mm deep, average 250mm finished thickness, level and to falls, cross falls or cambers

Base dimensions for
Fill to plan shape D

Filling to plan shape E
Existing G levels   6500
     5000
     4500
     6000
    4)22000
Av existg GL = 5500
Less Red Lvl reqd 6500
Av fill = (1000)

Figure 5.7 continued.

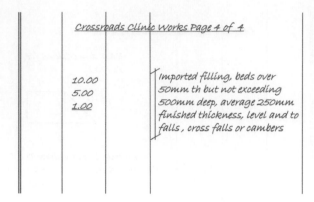

Figure 5.7 continued.

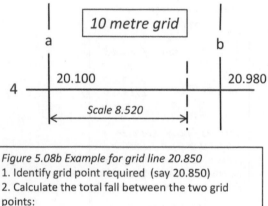

Figure 5.08b Method
1. Decide where the cut and fill contour line cuts through the grid.
2. Calculate the fall between these two grid points.
3. Calculate the fall between the point required (R.L.) and the lower grid point.
4. Divide the result of 3. by 2. and multiply the result by the size of the grid.
5. Scale or plot the answer from the lower grid point.
6. Repeat 1. to 5. for each grid line that is cut by the contour line.

Figure 5.08b Example for grid line 20.850
1. Identify grid point required (say 20.850)
2. Calculate the total fall between the two grid points:
       (B4) 20.980 – (A4) 20.100 = 0.880
3. Establish the difference the point required and the lower grid point:
       20.850 – (A4) 20.100 = 0.750
4. <u>Point required</u> x grid size
     Total fall
       <u>0.750</u> x 10.000 = 8.520 m
       0.850
5. Scale answer 8.520m from the lower grid point

Figure 5.8 How to plot cut-and-fill lines.

Few cut-and-fill excavations are as straightforward as the above example. Frequently the positioning of a cut-and-fill line will require other considerations, such as the removal of topsoil and excavating and filling in-depth ranges. It is very unlikely that a cut-and-fill line will fall consistently at the midpoint between two whole-number grid points. In a case where positioning a cut-and-fill contour line is less obvious, the following approach can be adopted (see Figure 5.8).

1   Decide where the cut-and-fill contour line cuts through the grid.
2   Calculate the fall between these two grid points.
3   Calculate the fall between the point required (R.L.) and the lower grid point.
4   Divide the result of (3) by (2) and multiply that by the size of the grid.
5   Scale or plot the answer from the lower grid point.
6   Repeat (1–5) for each grid line that is cut by the contour line.

5.02.04 Trench fill and strip foundations

Once the bulk (surface excavation) has been completed, the measurement of foundation excavation (trenches, pits, etc.) can commence. The starting level for this excavation is determined by the reduced level. Where the commencing level is not the original ground level, the level must be stated in the description.

5.02.04.01 Trench fill

The simplest (and some would argue, the cheapest) form of substructure is a trench fill foundation (Figure 5.9).

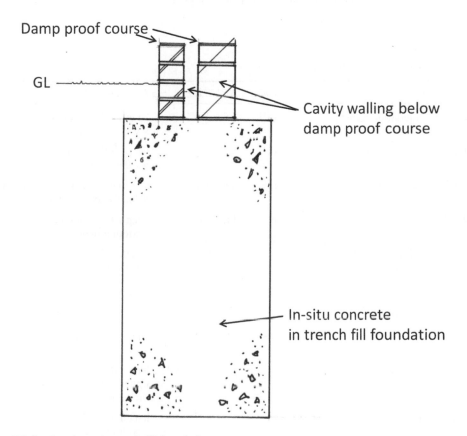

Figure 5.9 Section through trench fill foundation.

It is also a useful starting point for introducing the measurement techniques normally associated with foundation work. Table 5.2 provides a take-off list together with an indication of the unit of measurement, the NRM2 reference and an appropriate set of base dimensions for a trench fill foundation.

Assuming the site has already been stripped of topsoil and that any reduce level excavation has been completed, the following approach (as shown in Table 5.2) can be adopted. If the construction work includes internal wall foundations it would be normal to measure each in turn, taking care to note the width of internal foundations since these are often narrower than their external wall equivalent.

Few foundations are as straightforward as the above take-off list. In practice, this means that any or all of the following may be included, depending on the circumstances (Table 5.3).

5.02.04.02 Strip foundations

Strip foundations differ from trench fill foundations in that a strip of concrete is placed in the bottom of the trench which, in conjunction with masonry work and backfilling material, forms the basis of support for the superstructure. The approach to measurement is, in most respects, identical to a trench fill foundation and differs only in the measurement of backfill material. Figures 5.10 and 5.11 show backfill to a strip foundation.

Table 5.2 Take-off list for trench fill foundation.

Item	Unit	NRM2 Ref	Base dimensions
1. Excavate foundation trench	m^3	5.6.2.1-3	length (CL of founds) × width × depth
2. Disposal excavated mats	m^3	5.9.2.1-3	length (CL of founds) × width × depth
3. Earthwork support (only measured where specified)	m^2	5.8.1.1.1	(ext trench line × trench height) + (int trench line × trench height)
4. Blinding concrete	m^3	11.1.1.1	length (CL of founds) × width × depth
5. Concrete in foundation trench (horizontal work)	m^3	11.2.1-2.2.1	length (CL) × width × height
6. Masonry work to dpc (Cavity work – each skin measured separately on own Centre Line)	m^2	14.1.1.1	length (CL) × height
7. Cavity fill	m^3	11.5.1.1	length (CL of cavity × width × height
8. Damp Proof Course*	m^2	14.16-17.1-3.3	length (CL of wall) × width of walling. *If cavity work each dpc should be measured separately
7. Topsoil backfill	m^3	11.1.1-2.1	length (CL of fill) × width × height

Table 5.3 Take-off list trench fill foundation (potential measurable items).

Situation	Unit	NRM2 Ref	Treatment
1. Hazardous material			
2. Non-hazardous material			measure *extra-over* work previously measured
3. Excavating below ground water level*	m³	5.7.1.1-5	
4. Excavating in running water			
5. Unstable ground			
6. Excavating next to existing services	m	5.7.3.1.1	measure *extra-over* work previously measured
7. Excavating around existing services	nr	5.7.4.1.1	
8. Excavating in existing hard materials met in excavation	m³	5.7.1.1-5	measure *extra-over* work previously measured
9. Breaking up existing hard materials (rock, reinforce concrete, concrete, masonry or stonework)	m³	5.7.2.1-4	measure *extra-over* work previously measured

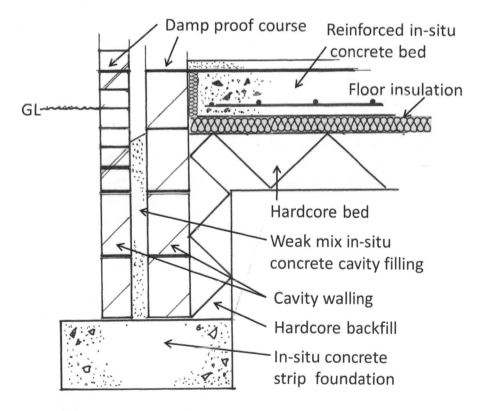

Figure 5.10 Section through strip foundation.

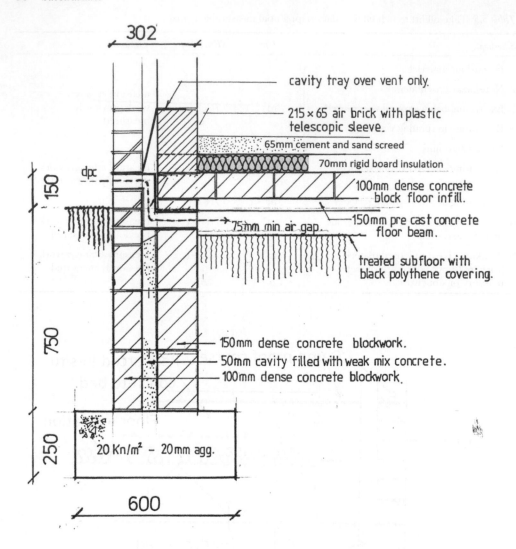

302

cavity tray over vent only.

215 × 65 air brick with plastic telescopic sleeve.

65mm cement and sand screed

70mm rigid board insulation

100mm dense concrete block floor infill.

150mm pre cast concrete floor beam.

75mm min air gap.

treated subfloor with black polythene covering.

dpc

150

750

150mm dense concrete blockwork.

50mm cavity filled with weak mix concrete.

100mm dense concrete blockwork.

250

20 Kn/m² – 20mm agg.

600

Figure 5.11 Section through independent ground floor construction (beam and block infill floor).

This should be measured as filling (NRM2 5.11.1–2.1–2.1), in square metres where the depth of fill does not exceed 500 mm depth and in cubic metres where it does. The material used and the method of backfilling will depend on the substructure construction, but it is worth remembering that backfilling with excavated material below a solid ground floor should be avoided and that backfilling with satisfactorily compacted hardcore would only be acceptable where the fill material is less than 600 mm thick or where the floor slab design includes reinforcement.

Bearing the above in mind, the following assumes an independent ground floor construction with floor loading passing to external and internal walls (ref section on previous page). The following is traditionally recognised as the approach to be adopted when booking dimensions where backfilling with excavated materials is required; the notes and drawings are intended to help understand the principles used. The example below (Figure 5.12) assumes a trench centre line of 36.00 m; that backfill material is to be obtained from trench excavation; and that any surplus excavated material will be removed from site. Other items that would normally be measured have been omitted.

The stages for establishing the quantity of filling and disposable excavated material are as follows.

Stage 1 Book the volume of trench excavation together with an equal volume of filling. The filling measurement is a theoretical 'paper one'. The trench will not be completely backfilled on site until the structural walls and strip foundations are in place.

Stage 2 Concrete in the foundation trench is measured; which will displace the fill material covered by the theoretical Stage 1 measurement. The now partly redundant fill material is no longer required and can be measured as disposal of excavated material.

Stage 3 Along the same lines as Stage 2, masonry work is measured in the foundation trench, which will once again displace the theoretical fill material previously booked. Similarly, part of the fill material is no longer required and can also be measured as disposal of excavated material. However, the masonry work will be measured up to dpc level and in square metres. The adjustment of the filling materials requires these to be made up only to ground level, but must be converted to cubic metres.

At this stage it is appropriate to establish the actual quantities of excavation, disposal and backfill; this can be carried out by preparing a mini-abstract.

The above example demonstrates the traditionally recognised procedure for booking dimensions where excavated material is backfilled around foundations. As already mentioned, there are occasions when this technique is inappropriate. Current building practice is more likely to require a detail similar to that shown in Figure 5.13.

In this example (Figure 5.13), imported hardcore must be used as backfill below a structurally dependent slab. The backfill material used externally differs from the material used internally and is of irregular shape. In this situation it would be more appropriate to adopt the following approach:

1   Excavate foundation trench.
2   Dispose of all excavated materials off site.
3   Find centre line of hardcore backfill and book volume.[*]
4   Find centre line of excavated material backfill and book volume.
5   Deduct the same volume as in (4) from material previously disposed off site (this can be added to item (4)).
6   Add the same volume as in (4), but this time describe this as 'disposed of on site in temporary spoil heaps'.

[*] The irregular sectional shape (parallelogram) of hardcore backfill must be booked as a volume and can be broken down into two simpler shapes (a rectangle and a triangle), with each booked as a volume on their individual centre lines.

5.03 Ground floors

To complete the work that would normally be included as part of substructure measurement, we must now consider the measurement of ground floors. For most residential low-rise constructions, the alternatives can be classified as follows:

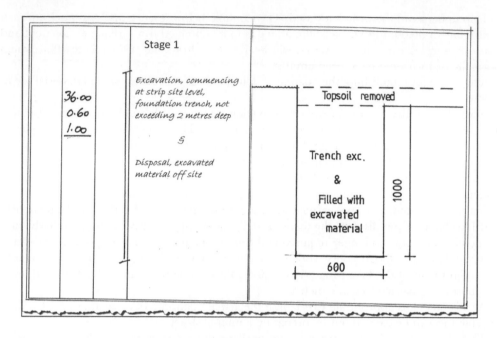

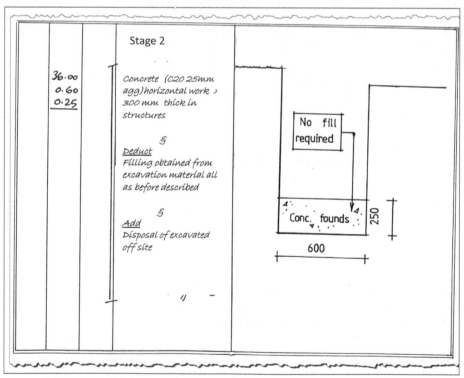

Figure 5.12 Traditional approach for booking dimensions where backfilling with excavated materials is appropriate.

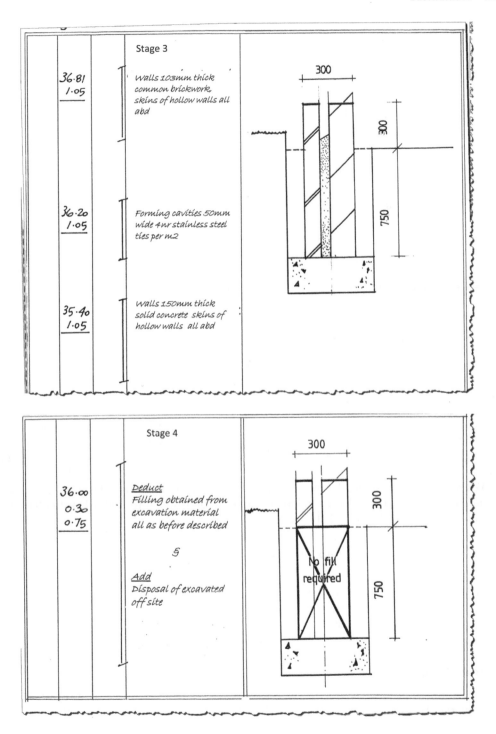

Figure 5.12 continued.

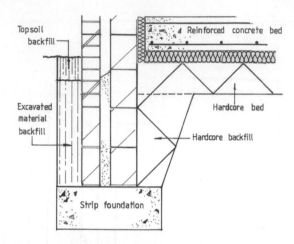

STRIP FOUNDATION AND BACKFILL

Figure 5.13 Section through strip foundation showing alternative backfill material (internal v external).

1   Suspended timber (partially independent).
2   Solid slab (dependent).
3   Concrete beam and block infill (independent).

5.03.01 Suspended timber ground floors

Popular in the past, their use today is infrequent. As the title suggests, this method of ground floor construction is independent of the strata immediately below the floor. Ground floor loads are transmitted via external walls to the foundations, or where spans are excessive, by the inclusion of sleeper walls (Figure 5.14).

Where fill is incorporated below a ground floor that is more than 600 mm deep, it is recommended that a suspended or independent floor construction is used. Measurement will usually include the following (Table 5.4).

5.03.02 Solid concrete ground floor slabs

This floor type bears directly onto the ground or filling layer immediately below, with ground floor loads transmitted directly through the fill material. The introduction of mesh (fabric) reinforcement to the slabs, together with the design of ground floor slabs that bear on the inner skin of the cavity wall, will reduce the risk of slab settlement (see Figure 5.13).

Measurement would usually include the following (Table 5.5).

5.03.03 Concrete beam and block infill floors

These have become increasingly popular as a method of avoiding problems of settlement caused by solid slab floors. The system works on the same principle as a suspended timber floor, by transmitting floor loads via structural floor beams to external and internal load-bearing walls. The beams are spaced at intervals that correspond with standard concrete block sizes, allowing insulated blocks laid flat, topped with rigid board insulation to complete the floor surface.

SUSPENDED TIMBER JOISTED FLOOR
(Independent)

SUSPENDED TIMBER JOISTED FLOOR
(Dependent)

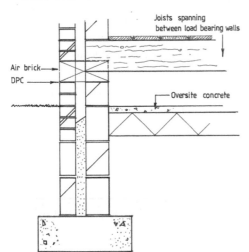

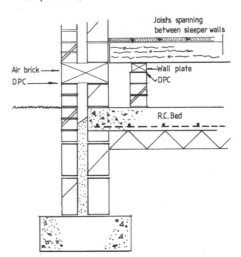

Figure 5.14 Suspended joisted floors (a) independent, (b) dependent.

Table 5.4 Take-off list for suspended joisted floor.

Item	Unit	NRM2 Ref	Base dimensions
1. Filling to make up levels	m³	5.12.2/3.1.1	plan length × plan width × depth
2. Concrete bed	m³	11.2.1.2.1	plan length × plan width × depth
3. Reinforcement (mesh)	m²	11.37.1/2/3	plan length (less cover) × plan width (less over)* (where sleeper walls bear on slab)
4. Sleeper walls	m²	14.1.*.1	length × height (Honeycomb bond)
5. Damp proof course	m²	14.16.1.3	length × width
6. Air bricks	nr	14.25.1.1.1	
7. Wall plate	m	16.1.1.3	length
8. Floor joist	m	16.1.1.4	length
9. Floor boarding	m²	16.4.2.1	plan length × plan width

5.04 Groundwater

As the name suggests, 'groundwater' is water that is present in the ground. The highest level of groundwater is more commonly referred to as the top of the water table. This may seem obvious, but it is necessary when undertaking substructure work to distinguish groundwater from surface water. Surface water is water that would normally fall as rain and/or collect on the surface of the ground. In regard to NRM2, the former is a measurable item and the latter is deemed included.

The presence and subsequent measurement of groundwater (or subsoil water) in foundations is determined by establishing (normally as part of the site investigation report) the depth (or

Table 5.5 Take-off list for in situ concrete ground floor.

Item	Unit	NRM2 Ref	Base dimensions
1. Filling to make up levels (* not to exceed 500 mm thick)	m³	5.12.2.1	plan length × plan width × depth
2. Concrete bed	m³	11.2.1.2.1	plan length × plan width × depth
3. Formwork (if necessary)	m	11.13.1	perimeter length
4. Reinforcement (mesh)	m²	11.37.1/2/3	plan length (less cover) × plan width (less cover)
5. Board insulation (horizontal)	m²	31.1.1.1.1	plan length × plan width
6. Board insulation (vertical)	m²	31.1.1.1.2	Int wall perim × height
7. Damp proof membrane – (flexible sheet) – (liquid applied) – (mastic asphalt)	m² m² m²	19.1.1.4 19.1.1.4 19.1.1.4	length × width

Floor screeds do not usually form a structural element of floor construction and would be included as part of the floor finishes measurement (NRM2.28.1.1.1).

datum height) of the groundwater below existing ground level. This level is recorded and established as the 'pre-contract groundwater level'. Once excavation operations commence, the groundwater level is re-established and recorded as the 'post-contract water level'. Where these two levels differ, measurements are revised accordingly (NRM2 5.9.1.1.1).

Even though these levels may have changed between pre- and post-contract readings (groundwater levels are likely to change with the seasons), in many instances the groundwater level will remain below the lowest point of the foundations and there will be no need for an adjustment.

In a situation where groundwater levels would have been encountered as part of the intended substructure operation, or where groundwater levels have subsequently risen and affected the progress of substructure operations, measurements and costings are revised accordingly.

In this situation, in addition to the normal foundation excavation (NRM2 5.6.1/2.*.*) an item is measured as 'extra-over' the initial foundation measurement. This is given in cubic metres, but only to the depth that groundwater was recorded in the trench (i.e. to the height of the post-contract water level) (see Figure 5.15).

5.05 Ground remediation and soil stabilisation

5.05.01 Ground remediation

The New Rules of Measurement 2 (NRM2) include a Work Section to accommodate the measurement of preparatory works in order to make good contaminated ground. It is likely that remedial work would be necessary where any land had previously been used for manufacturing, agricultural or industrial purposes. In such cases the ground may contain toxins and other pollutants that are harmful to humans and will need to be treated, contained or removed from site before any construction work can commence. A soil investigation report should be commissioned in order to identify the existence of any contamination. The details from this report need to be provided as part of the mandatory information that accompanies the measured work (NRM2 Mandatory Information 6.1.3). It will also be necessary to provide ground conditions, groundwater levels and the dates that these were established, together with the starting levels for each type of excavation, all of which should be given as part of the Work Section mandatory information.

Some provision for remediation is also included in NRM2 work section 5, Excavation and Filling (NRM2 5.7.1.1). No specific guidance is offered as to the specific circumstance that dictates where one applies and the other does not. By way of general guidance, it would seem logical to invoke NRM2 work section 6 where contamination is extensive or evident over an entire site, while an isolated incident of contamination in an otherwise 'clean' site can be dealt with under NRM2 work section 5.7.1.1.

With the exception of site de-watering (NRM2 6.1), which is measured as an item, the predominant unit of measurement for ground remediation measurement is square metres. Items of work identified as requiring superficial measurement include sterilisation, chemical neutralising and freezing (NRM2 6.2, 6.3, 6.4). The detail of the description in each case should include the maximum depth of ground to be treated and the method proposed. Where the ground remediation technique adopted includes freezing, in addition to describing the method of freezing it will be necessary to give the duration of the freezing process where this is not left to the discretion of the contractor. Ground gas venting (NRM2 6.5) is also measured in square metres, with a description that identifies the type of gas to be vented and the method of collection and disposal.

5.05.02 Soil stabilisation

This is a term that describes techniques used to improve the performance of the substrata by enhancing the ability of the ground to receive loading. In so doing, a site that might otherwise not have been considered for development becomes viable. It is necessary to give the same set of mandatory information for soil stabilisation as was provided for ground remediation (NRM2 Mandatory Information 6.1.3) (see further mandatory information required under ground remediation). With the exception of ground anchors, which are enumerated (NRM2.6.7), soil nailing, pressure grouting/ground permeation, compacting and stabilising soil in situ are all measured in m² (NRM2 6.6, 6.8, 6.9, 6.10).

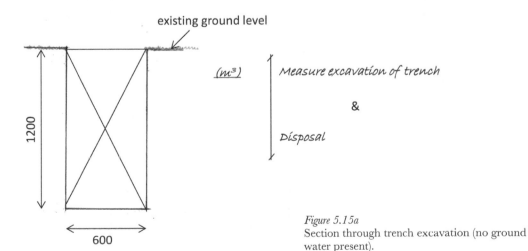

Figure 5.15a
Section through trench excavation (no ground water present).

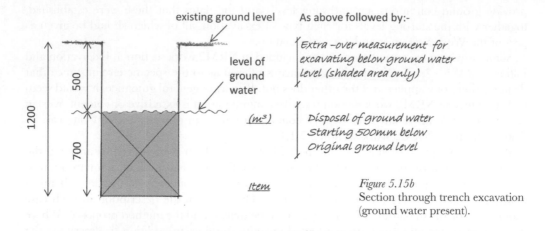

existing ground level

As above followed by:-

level of ground water

Extra –over measurement for excavating below ground water level (shaded area only)

(m^3)

Disposal of ground water starting 500mm below original ground level

Item

Figure 5.15b
Section through trench excavation (ground water present).

5.06 Trench fill foundation job ref.: ATR 200289

5.06.01 Plan, section, specification notes

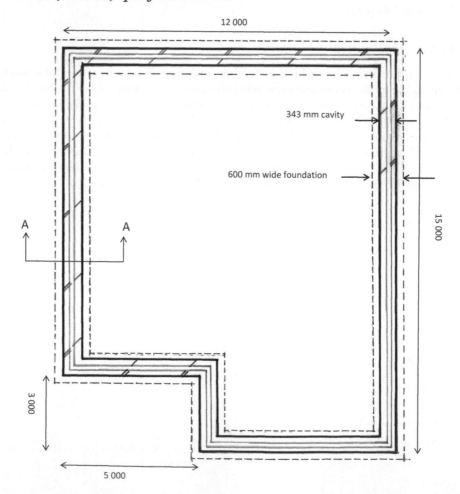

12 000

343 mm cavity

600 mm wide foundation

15 000

A   A

3 000

5 000

5.06.02 Measurement example

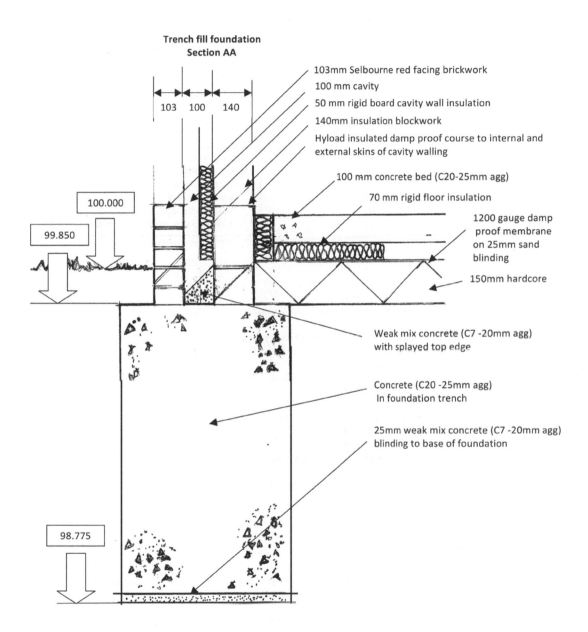

**Trench fill foundation
Section AA**

103 100 140

100.000

99.850

98.775

103mm Selbourne red facing brickwork

100 mm cavity

50 mm rigid board cavity wall insulation

140mm insulation blockwork

Hyload insulated damp proof course to internal and external skins of cavity walling

100 mm concrete bed (C20-25mm agg)

70 mm rigid floor insulation

1200 gauge damp proof membrane on 25mm sand blinding

150mm hardcore

Weak mix concrete (C7 -20mm agg) with splayed top edge

Concrete (C20 -25mm agg) In foundation trench

25mm weak mix concrete (C7 -20mm agg) blinding to base of foundation

5.07 Basement Job ref.: 12/GRELSA

Substructures – Trench Fill Foundation   Job ref: ATR 200289   Page 01 of 07

The work comprises the excavation and associated substructures for a detached residential
property all in accordance with Structural Engineer's drawing reference ATR200289
The ground is assumed to consist of 150mm of topsoil overlaying clay and sandstone
depths all as detailed in the Structural Engineer's borehole log dated 04/11/2020
The site is substantially level with existing GL established at 100.000
Groundwater level was established at 4.60m (95.400) below existing GL on 04/11/2020
No over or underground services cross the site

> Measurable Smm7 groundworks items
> deemed included under NRM2
> * Disposal of surface water
> * Levelling and compacting
> * Working space
> * Earthwork Support (discretionary)

Take off list

TRENCH FILL FOUNDATION
* Excavate topsoil
* Retain excavated materials on-site
* Foundation excavation
* Disposal excavated materials off-site
* Support to faces of excavation
* Blinding Concrete
* Mass concrete in foundations
* Brick walling (up to dpc)
* Forming cavities (ditto)
* Rigid board wall insulation
* Block walling (ditto)
* Damp proof course
* Concrete filling to hollow wall

CONCRETE OVERSITE (BED)
* Hardcore filling
* Sand blinding
* Damp proof membrane
* Board Insulation
* Concrete bed

Note:- Given the nature of the substrata the Structural Engineer has stipulated that
temporary support to the sides and faces of open excavation will be required.

It is assumed that the site has been cleared of all vegetation and clear contractor access is
possible.

<u>*Site Preparation*</u>

Topsoil excavation dimension

    Projection of founds
    beyond face of wall

Foundation width	600
Less walling width	<u>343</u>
projection both sides	257
proj one side 2)257 =	<u>128.5</u>

External wall plan dimension
         L   *W*
     15 000   12 000

Add
projection
2/128.5   <u>257</u>   <u>257</u>
        <u>15 257</u> <u>12 257</u>

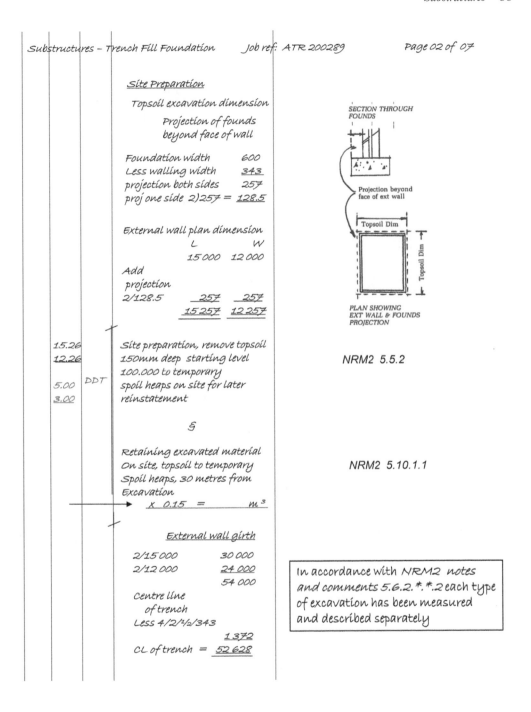

SECTION THROUGH FOUNDS

Projection beyond
face of ext wall

Topsoil Dim

Topsoil Dim

PLAN SHOWING
EXT WALL & FOUNDS
PROJECTION

15.26
12.26

5.00 DDT
3.00

Site preparation, remove topsoil
150mm deep starting level
100.000 to temporary
spoil heaps on site for later
reinstatement

NRM2 5.5.2

&

Retaining excavated material
On site, topsoil to temporary
Spoil heaps, 30 metres from
Excavation
      x 0.15 =   m³

NRM2 5.10.1.1

<u>*External wall girth*</u>

2/15 000	30 000
2/12 000	<u>24 000</u>
	54 000

Centre line
of trench
Less 4/2/½/343
            <u>1 372</u>
CL of trench = <u>52 628</u>

In accordance with *NRM2* notes
and comments 5.6.2.*.*.2 each type
of excavation has been measured
and described separately

Substructures – Trench Fill Foundation   Job ref: ATR 200289   Page 03 of 07

<u>Depth of trench</u>

	150
	1 050
	25
	1 225
Less topsoil	150
	1 075

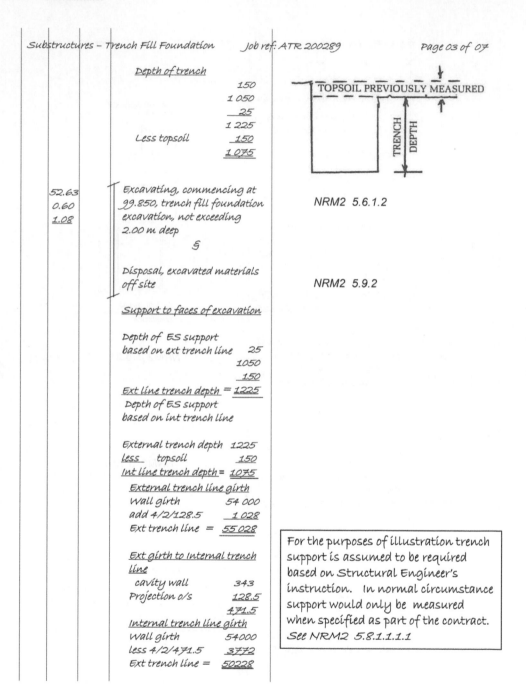

TOPSOIL PREVIOUSLY MEASURED

TRENCH DEPTH

52.63	Excavating, commencing at	NRM2 5.6.1.2
0.60	99.850, trench fill foundation	
1.08	excavation, not exceeding	
	2.00 m deep	
	&	
	Disposal, excavated materials	NRM2 5.9.2
	off site	

<u>Support to faces of excavation</u>

Depth of ES support
based on ext trench line   25
                                          1050
                                          150
<u>Ext line trench depth</u> = <u>1225</u>
Depth of ES support
based on int trench line

External trench depth 1225
<u>Less</u>   topsoil   150
<u>Int line trench depth</u> = <u>1075</u>
<u>External trench line girth</u>
Wall girth   54 000
add 4/2/128.5   1 028
Ext trench line = <u>55 028</u>

<u>Ext girth to Internal trench</u>
<u>line</u>
  cavity wall   343
Projection o/s   128.5
                              471.5
<u>Internal trench line girth</u>
Wall girth   54000
less 4/2/471.5   3772
Ext trench line = <u>50228</u>

For the purposes of illustration trench
support is assumed to be required
based on Structural Engineer's
instruction.   In normal circumstance
support would only be measured
when specified as part of the contract.
See NRM2 5.8.1.1.1.1

Substructures – Trench Fill Foundation   Job ref: ATR 200289   Page 04 of 07

Trench Support (contd)

55.03 1.23 50.23 1.08	Support to faces of excavation _{ext} where not at the discretion of the contractor, maximum (int depth 1225 mm, close boarded timber support

NRM2 5.8.1.1.1.1

52.63 0.60 0.05	Concrete Work in Substructures Plain in-situ blinding concrete (C7 - 20mm agg) horizontal work ≤ 300mm in structures, poured on our against earth or un-blinded hardcore.

NRM2 11.2.1.1.1

NRM2 11.2.2

Trench fill founds

52.63 0.60 1.08	Plain in-situ concrete (C20 – 25mm agg) horizontal work › 300 mm thick in structures

Cavity Walls up to dpc

Centre line of external cavity wall

Ext girth   54 000
Less 4/2/$\frac{1}{2}$/ 103   412
CL of ext skin   53 588

Centre line of cavity

ext wall   103
middle of cavity $\frac{100}{2}$ = $\frac{50}{153}$

Ext girth   54 000
Less   4/2/153   1 224
CL of cavity   52 776

Centre line of internal wall

ext wall   103
cavity   100
int wall $\frac{140}{2}$ = $\frac{70}{273}$

Ext girth   54 000
Less   4/2/273   2 184
CL of int skin   51 816

<u>Height of external walling to dpc</u>
assume five courses @ 75mm
Per course = 5 / 75 = <u>375mm</u>

53.59	Walls (103 mm thick) in skins of hollow walls in Selbourne red facing bricks laid stretcher bond in cement mortar (1:3) with a neat rubbed joint as work proceeds.	NRM2 14.1.1.1
0.38		

52.78	Forming cavity, 100mm wide including 4 nr stainless steel wall ties with insulation retaining clips per m²	NRM2 14.14.1.1
0.38		

§

Rigid board cavity wall Insulation Celotex CG5000, 50mm thick, plain areas, horizontal fixed with plastic retaining clips awp.

NRM2 31.1.1.1.1

> Cavity wall insulation included for convenience with the cavity forming measurement but could also be measured separately on own centre line with superstructure cavity wall

<u>Average height of cavity filling</u>

Max   2/75   150mm
Min   <u>75mm</u>
                2) 225 mm
Average   = <u>175 mm</u>

52.78	Weak mix concrete (C7 - 20mm agg) in Cavity fill vertical work ≤ 300mm thick	NRM2 11.5..1
0.10		
0.18		

51.82	Walls (140 mm thick) insulation blockwork, skins of hollow walls laid stretcher bond in cement mortar (1:3) with a neat rubbed joint as work proceeds.	NRM2 14.1.2.1
0.38		

Damp proof courses

<u>53.59</u> <u>51.82</u>		Hyload insulated damp (ext wall proof courses ≤ 300 mm , (int wall bedded in cement mortar (1:3) horizontal.	NRM2 14.16.1.3

Oversite and gf Concrete bed

External wall plan dimension

	L	W
	15 000	12 000

<u>Less</u>
cavity wall

2/343	<u>686</u>	<u>686</u>
	<u>14314</u>	<u>11314</u>

14.31 <u>11.31</u> 5.00 <u>3.00</u>	DDT	Hardcore (MOT Type 2) imported filling in beds over 50 mm but not exceeding 500mm deep, finished thickness 150mm, level to falls, cross falls or cambers, in sub base to concrete bed	NRM2 12.2.1.1

&

Imported filling, sand in
blinding bed not exceeding
50mm thick, finished thickness
25mm, level to falls,
cross falls or cambers, as
blinding layer to dpm

NRM2 12.1.1.1

Rigid floor insulation (horiz)
and concrete bed

External wall plan dimension

	L	W
	15 000	12 000

cavity wall 343
Vertical ins 70
             413
Less 2/413   826   826
              14174   11174

14.17		Rigid floor insulation, boards	NRM2 31.1.1.1
11.17		70mm thick, plain areas,	
	DDT	horizontal	
5.00			
3.00		&	

Plain In-situ concrete in beds
(C20 -25 mm agg.) horizontal   NRM 2 11.2.1
work ≤ 300mm thick, in
structures
→   X 0.30 =   m3

Rigid floor insulation (vertical)

Ext girth   54 000
 Less 4/2/343   2 744
internal wall girth   51 256

Height of vertical
insulation
  horiz insulation th   70
  floor slab th   100
                         170

51.26	Rigid floor insulation, boards	NRM2 31.1.1.2
0.17	70mm thick, plain areas,	
	vertical	

5.07.01 Plan, section, specification notes

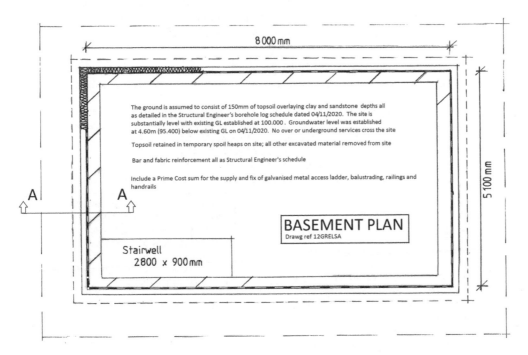

The ground is assumed to consist of 150mm of topsoil overlaying clay and sandstone depths all as detailed in the Structural Engineer's borehole log schedule dated 04/11/2020. The site is substantially level with existing GL established at 100.000. Groundwater level was established at 4.60m (95.400) below existing GL on 04/11/2020. No over or underground services cross the site

Topsoil retained in temporary spoil heaps on site; all other excavated material removed from site

Bar and fabric reinforcement all as Structural Engineer's schedule

Include a Prime Cost sum for the supply and fix of galvanised metal access ladder, balustrading, railings and handrails

BASEMENT PLAN
Drawg ref 12GRELSA

8 000 mm

5 100 mm

Stairwell
2800 x 900 mm

A   A

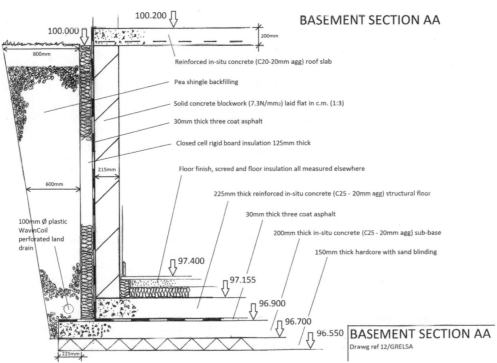

BASEMENT SECTION AA

100.200

100.000

800mm

215mm

600mm

100mm Ø plastic WavinCoil perforated land drain

200mm

Reinforced in-situ concrete (C20-20mm agg) roof slab

Pea shingle backfilling

Solid concrete blockwork (7.3N/mm²) laid flat in c.m. (1:3)

30mm thick three coat asphalt

Closed cell rigid board insulation 125mm thick

Floor finish, screed and floor insulation all measured elsewhere

225mm thick reinforced in-situ concrete (C25 - 20mm agg) structural floor

30mm thick three coat asphalt

200mm thick in-situ concrete (C25 - 20mm agg) sub-base

150mm thick hardcore with sand blinding

97.400

97.155

96.900

96.700

96.550

225mm

BASEMENT SECTION AA
Drawg ref 12/GRELSA

5.07.02 *Measurement example*

See also measurement example in Chapter 8 (Structural Steel Frame) for work associated with excavating and placing pad foundations; and Chapter 15 (External Works) for work associated with roadways, paving, planting trees and laying turf.

Substructures – Basements Job ref: 12/GRELSA   Page 01 of 08

The work comprises the excavation and associated substructures for a stand-alone basement all in accordance with the Structural Engineer's drawing reference 12/GRELSA. The ground is assumed to consist of 150mm of topsoil overlaying clay and sandstone depths all as detailed in the Structural Engineer's borehole log dated 30/06/2020. The site is substantially level with existing GL established at 100.000. Groundwater level was established at 4.60m (95.400) below existing GL on 04/11/2020. No over or underground services cross the site

> Measurable Smm7 Groundworks items deemed included under NRM2
> - Disposal of surface water
> - Levelling and compacting
> - Working space
> - Earthwork Support (discretionary)
> - Angle fillets/edges

Take off list

BASEMENT CONSTRUCTION
- Topsoil excavation
- Retain excavated materials on-site
- Bulk excavation
- Disposal excavated materials off-site
- Hardcore filling
- Concrete sub base
- Asphalt tanking (horizontal)
- Reinforced concrete base
- Formwork to sides of foundation
- Reinforcement from schedule
- 215mm block walling
- Asphalt tanking (vertical)
- Asphalt apron
- Insulation (vertical) to block walling
- Concrete bed
- Reinforced concrete roof slab
- Reinforcement from schedule
- Adjustment for stairwell
- Formwork to edges of horizontal work
- Formwork to soffits of slab
- PC Sum for galvanised metal access ladder
- Adjust disposal of excavated materials
- Land drainage to perimeter of basement
- Pea shingle backfill around basement

Notes:-
Given the nature of the substrata the Structural Engineer has made no stipulation that temporary support to the sides and faces of open excavation will be required.

It is assumed that the site has been cleared of all vegetation and uninhibited contractor access is possible.

Floor insulation, screed and floor finish measured elsewhere

All reinforcement measured directly from Structural Engineer's reinforcement schedule

<u>Site Preparation</u>

Topsoil excavation dimension

Projection of founds
 beyond face of wall
  asphalt tanking   30
  insulation   125
 Max proj for gran fillg 800
                         955

<u>External wall plan dimension</u>

	L	W
	8 000	5 100
Add projection		
2/955	1 910	1 910
	9 910	7 010

9.91 7.01	Site preparation, remove topsoil 150mm deep starting at existing ground level to temporary spoil heaps on site for later reinstatement

NRM2 5.5.2

&

Retaining excavated material on site, topsoil to temporary spoil heaps, 30 metres from excavation

NRM2 5.10.1

x 0.15 =   m³

> In accordance with NRM2 6.2.*.*.2 each type of excavation has been measured and described separately

<u>Depth of basement</u>

Existing GL datum 100.000
underside of hdcore 96.550
                      3 450
 <u>Less</u> topsoil   150
                      3 330

<u>Average projection width of</u>
granular back filling beyond
external basement wall

Width @ base

asphalt	30
insulation	125
projecting toe	225
Granular surround	<u>100</u>
width @ base	480
width @ GL	<u>800</u>
2)	<u>1 280</u>
<u>Av projection width</u> =	<u>640</u>

	L	W
	8 000	5 100
add 2/640	<u>1 280</u>	<u>1 280</u>
	<u>9 280</u>	<u>6 380</u>

9.28	Excavating, commencing at
6.38	strip site level bulk excavation,
<u>3.30</u>	over 2.00 m deep not
	exceeding 4.00 m deep

&

Disposal, excavated materials
off site

<u>Hardcore dimensions</u>

Projection beyond block wall

asphalt	30
insulation	125
projecting toe	<u>225</u>
	<u>380</u>

	L	W
	8 000	5 100
add 2/380	<u>760</u>	<u>760</u>
	<u>8 760</u>	<u>5 860</u>

8.76 5.86	Imported hardcore filling in beds exceeding 50mm thick ≤ 500mm deep level, 150mm thick with sand blinding
	NRM2 5.6..1

$$X \quad 0.15 \quad = \qquad m^3$$

NRM2 5.9.2.1

&

Plain in-situ concrete (C25 - 20mm agg) ≤ 300mm in structures

&   NRM2 11.2.1.1.1

$$X \quad 0.20 \quad = \qquad m^3$$

Asphalt tanking coverings >500mm wide, horizontal protection to concrete base 30mm thick, 3 coat application

NRM2 11.2.1.2.

Basement structural slab

8.00 5.10 0.23	Reinforced in-situ concrete (C25-20mm aggregate) horizontal work ≤ 300mm in structures

(Note reinforcement measured directly from schedule).

> The Structural Engineer has not Stipulated that trench support is required. In normal circumstance support would only be measured when specified as part of the contract. See NRM2 5.8.1.1.1.1

	<u>Formwork to edge of bsmt slab</u>	
	2/ 8 000	16 000
	2/ 5100	<u>10 200</u>
		<u>26 200</u>
26.20	Formwork to sides of foundations and bases ≤ 500mm high; 225mm high	
	<u>Centre-line of basement wall</u>	
	External perim length of walling	
	2/ 8 000	16 000
	2/ 5100	<u>10 200</u>
	Ext wall girth	26 200
	Less 4/ 2/ 1/2 / 215	<u>860</u>
	CL of block wall	<u>25 340</u>
	<u>Height of blockwork</u>	100.000
	less	<u>97.155</u>
		2.845
25.34 2.85	Walls (215mm thick) solid concrete blocks 7.3N laid flat in stretcher bond in c.m. (1:3) joints raked one side to provide key for asphalt tanking	
	<u>vertical asphalt tanking</u>	
	datum @ top of slab	100.200
	less	<u>96.900</u>
		<u>3 300</u>
	girth of asphalt as previous	<u>26 200</u>

26.20 3.30		Asphalt tanking coverings >500mm wide, vertical protection to blockwork base 30mm thick, 3 coat application
26.20		Asphalt abd apron 200mm girth horizontal

Basement wall insulation
height datum top of insulation

$$
\begin{array}{r}
100.000 \\
\underline{96.900} \\
\underline{3.070}
\end{array}
$$

girth of ext face of asphalt

$$26\,200$$

add $4/{2}/30$

$$
\begin{array}{r}
\underline{240} \\
\underline{26\,440}
\end{array}
$$

26.44 3.07		Closed cell rigid insulation board 125mm thick plain vertical areas
8.00 5.10 0.20	Ddt	Reinforced in-situ concrete (C20- 20mm aggregate) horizontal work ≤ 300mm thick in structures
2.80 0.90 0.20		(stairwell)

(note reinforcement measured
directly from schedule)

NRM2 11.2.1.2
Opening = 0.504m^3 therefore
concrete adjustment required

Girth of formwork to sides of
basement roof

  as previous = 26 200

26.20   Formwork to edges of horizontal
Work ≤ 500mm high

Formwork to soffit of basement
roof

	length	width
	8 000	5 100
less 2/		
215	430	430
	7 570	4 670

7.57
4.67   Formwork to soffits of
horizontal work for
concrete ≤ 300mm thick,
propping ≤ 3m high

No adjustment required to formwork
for stairwell void since opening area
is ≤ 5.00m2

Item   Prime Cost sum for the supply
and fix of galvanised metal
access ladder/stairway
including balustrading
railings and handrails

Perimeter land drain
(assuming pipe is central to
  basement toe projection)
$\frac{225}{2}$ = 112.5 say 113mm

asphalt	30
insulation	125
half toe projection	113
	268

<u>Perimeter land drain</u> (Contd)

girth of ext wall (as previous)

		26 200
add 4/2/268		2 144
		28 344

28.34 | Drain runs including bed and surround (excavation measured elsewhere) average depth 2 500 – 3 000mm; 100mm Ø plastic WavinCoil perforated land drain

Backfilling around basement and land drain

Depth of b'fill		100.000
	less	96.700
		3.300

Average width of b'fill
       as before   (P1)   640mm

External wall girth		26 200

<u>add</u> 30
     125
640 320
2   <u>475</u>

<u>Add</u> 4/2/1
        2/475/

	1 900
	28 100

28.10
0.64
3.30 | Imported pea shingle filling, beds exceeding 500mm deep as backfilling around basement construction and land drain

To take/check:-
• All Reinforcement

6   Masonry work

6.01 Introduction

In previous editions of the Standard Method, the term 'Masonry Work' had been used exclusively to describe the trade of the stone mason together with the dressing, coursing and laying of stone. NRM2 adopts this same term to embrace the trades of brickwork and blockwork together with artificial (cast) and natural stonework.

These are classified as follows.

NRM2 14 Masonry	14.1.1	Brickwork
	14.1.2	Blockwork
	14.1.3	Glass blockwork
	14.1.4	Natural stone
	14.1.5	Cast stone
	14.1.6	Other: type stated

When measuring brick and block walling it is always advisable to follow a set order (this would normally replicate the sequence of construction) and to provide a take-off list. Under normal circumstances masonry below damp proof course level would be included as part of the substructures measurement. There is no specific requirement in NRM2 to distinguish between work in substructures and work in superstructures. Nevertheless, the general approach to measurement would normally require that work in substructure and work in superstructure measurement are kept separate. On large projects where there are a number of different types of walling, it is helpful to colour code the drawings accordingly. The alternatives for different types of walling will be determined by location (substructure/superstructure, internal/external) or function (structural/non-structural).

The primary unit of measurement for Masonry Work is square metres. The two dimensions needed to provide the area are the centre-line length and the height of the walling. In order to price the measured masonry work, the measurer will also need to provide a description that identifies the walling thickness, the type of brick/block, the bond, the type of mortar and pointing (Figure 6.1; see NRM2 Work Section 14; Mandatory Information).

Rather than repeat these details in every description, a heading can be established at the commencement of masonry measurement to include these details. The subsequent descriptions then need only make reference to the wall's thickness, its plane (when not vertical) and whether it is built in facework on one or both sides. The thickness of walling given in the description can be recorded in one of two ways: either by reference to a standard brick size or by stating the sectional width of the walling. To avoid any doubt, some surveyors prefer to give both. (Refer to brick/block masonry descriptions and Figures 6.1a and b.)

The majority of cavity walls are constructed with an outer leaf in the brickwork, a cavity (which may or may not be insulated) and an inner leaf of blockwork (usually with some insulation properties). Each of these 'skins' of masonry work forms what is termed a 'hollow wall', and each component will require measuring as a separate item based on its own centre line. The centre line of a wall is based on its midpoint length and will require a waste calculation to establish this (see Chapter 3.03 and NRM2.14 Notes Comments and Glossary). The purpose and location of walls will largely determine the type of brick or block used and the finish on the face of the exposed wall. The following terms are commonly employed in both the construction industry and in NRM2 to distinguish between the various finishes of brickwork (Figure 6.2).

6.02 Brick/block walling descriptions

Since the cost of *facing bricks* can be significantly more than *common bricks*, it is necessary in the descriptive part of the measurement to distinguish between these two basic types of brick. The industry often refers to these different types of walling by adopting the following terminology when referring to a finished brick-built masonry wall.

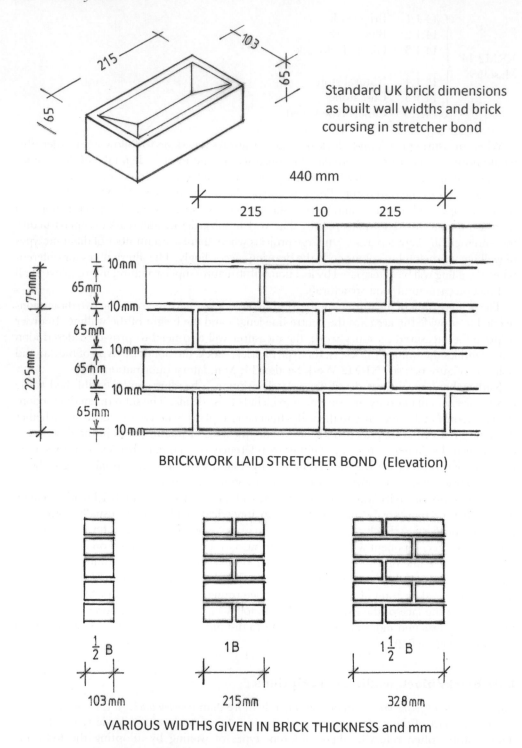

Figure 6.1 (a) Standard UK brick dimensions as built wall widths and brick coursing in stretcher bond.

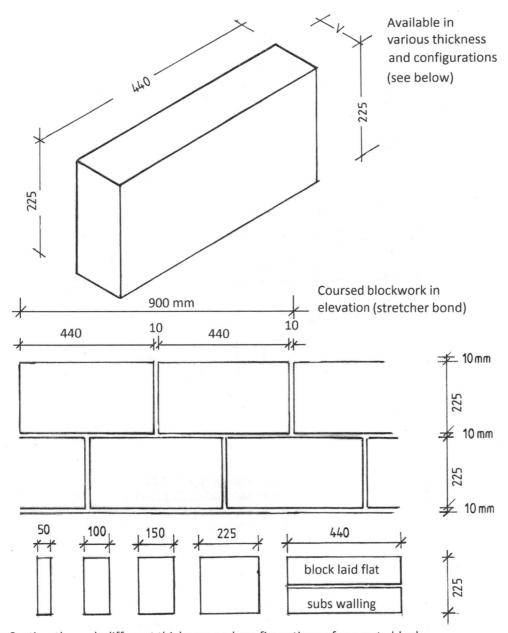

Available in various thickness and configurations (see below)

Coursed blockwork in elevation (stretcher bond)

Section through different thickness and configurations of concrete blocks

Figure 6.1 (b) Standard block dimensions.

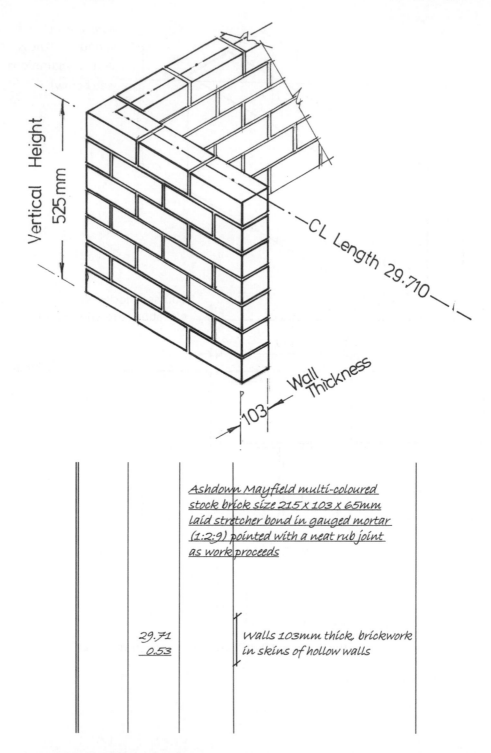

Figure 6.2 Brickwork (Masonry) measured in square metres (m2) based on the centre line of the walling.

6.02.01 Common brickwork

This is a wall built of any type of brick that has a structural function only and no particular attention is given to the finish. Where the choice of brick does not matter, the cheapest brick available will be used, and this is likely to be a common brick. Walls built of common brickwork will normally remain unseen once building work is completed. They are often used below ground level or in situations where they will subsequently be covered (Figure 6.3).

6.02.02 Facework

Facework one side or *facework both sides* indicate that the walling requires attention as to its final appearance, either to one face or both. This last term should not be confused with *facing bricks* or *facing brickwork*.

6.02.03 Facing brickwork

A particular selection of brick will have been made. Facing brickwork is walling built where the exposed or 'facing' part of the brick is left with a pleasing finish. By definition, this type of walling is finished 'facework'. In other words, care would be taken to avoid using chipped or irregular-shaped bricks and the mortar joints would be carefully pointed on completion.

There are six main classifications for brick/block walling (NRM2 14.1–6) and a host of other associated classes, ranging from isolated piers and projections to flues and flue linings (NRM2 14.4–12). All walling must be allocated for descriptive purposes to one of these classes. Mention should also be made of the shape or plane that the finished walling will take. All walling is presumed (deemed) to be vertical, battering (sloping walls with parallel sides) or tapering on one or both sides (walls of diminishing thickness). Where masonry is built against other work or used as formwork, this should be stated.

Either as part of the normal description, or perhaps as a general heading to the masonry brick/block walling trade, details must be provided of the Mandatory Information identified in NRM2 14 Mandatory Information 1–8. As noted previously, if the latter approach is adopted this inclusion will prevent unnecessary repetition. As is the case with all trades in NRM2, it will be necessary to identify the scope of the work in accordance with the requirements of the information offered at the start of this work section.

An alternative approach could be adopted where a heading is established on the sheet of dimension paper. This would save time where there are a number of descriptions relating to the same facing brickwork (Figures 6.4 and 6.5).

A similar procedure can be employed when measuring block walling. NB: In Figures 6.3, 6.4 and 6.5 the dimensions are assumed and waste calculations have been omitted.

6.03 Approach to measurement

6.03.01 Cavity walls

Each skin of a cavity wall is measured separately in square metres on a centre line. Forming cavities in hollow walls are measured in the same way, stating the width of the cavity in the description together with the type, size and spacing of wall ties (NRM2 14.14.1.1). If the cavity includes any rigid sheet insulation, this is measured in square metres, stating the type of

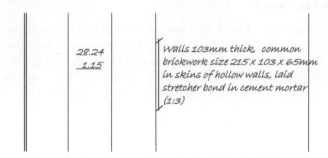

Figure 6.3　The brickwork description here suggests that this work will remain unseen (either in foundations or subsequently covered with plasterwork) since it is formed of common bricks and no detail is given in the description of facework. So this particular wall will be 'left as laid'.

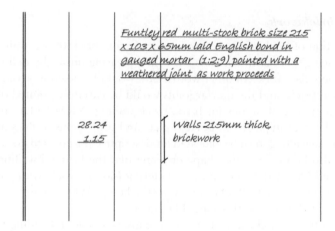

Figure 6.4　An example of setting up a heading on dimension paper to establish a common set of specification details that will apply to all of the following measured brickwork. In this particular instance the brickwork will be on view since the description makes reference to 'facework' and the brick being used is a facing brick.

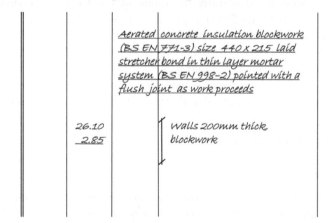

Figure 6.5　In this case a description heading has been set up for blockwork. The specification suggests that this walling forms the internal skin of a traditional cavity.

insulation and thickness along with the method of fixing or installation. It is presumed that both the cavity and insulation will be based on their respective centre-line measurement.

Openings in walls for windows and external doors are, at this stage, ignored. An adjustment will be made at the time of measuring the window or door, and the initial over-measurement reduced accordingly; see measured example Chapter 11 (Opening adjustments for windows page 3 of 7).

When measuring the external walling to traditional gable-ended properties, the stages for the measurement of masonry work would be as follows (Figure 6.6):

Stage 1 Measure a box (or rectangle) based on the centre line × height to soffit level, for each skin of the wall and the cavity.
Stage 2 At the eaves, in order to close the cavity along the length of the soffits, measure in linear metres along the length of the eaves as extra-over the initial superficial wall measurement (NRM2 14.11.1.2). Alternatively, the cavity wall may be left open at eaves level and the blockwork inner skin raised independently to support the wall plate (details will vary depending on circumstances). The wall plate will be measured as part of the Carpentry Work Section (NRM2 16).
Stage 3 Before measuring the triangular section of the gable end cavity walling at either end of the dwelling, it will be necessary to include a narrow band of cavity work to lift these end sections to the same height as the brickwork that was measured to close the cavities at soffit level (Stage 2).

6.03.02 Internal partition walls

To complete the measurement of the masonry work associated with low-rise residential property, it will be necessary to book dimensions for internal partition walls. Care should be taken to distinguish between stud walling and other internal walling systems, which must be measured under a different set of rules. In normal circumstances, the floor-to-ceiling height will be consistent on each storey and one of the following techniques may be adopted to facilitate the measurement process (Figure 6.7).

6.03.03 Masonry work (other classifications)

The previous sections have identified how to set down dimensions and descriptions for masonry work. To complete the measurement of this work it is necessary to explain a number of other categories associated with brickwork/blockwork. For ease of reference, these are given in accordance with the relevant NRM2 work section classification ref 14.4 to 14.12.

6.03.03.01 Attached piers and projections

When measuring attached piers, the height of the projection is booked in linear metres stating the width and depth together with the plane in the description. This should be accompanied by a dimensioned description or a dimensioned diagram. Isolated piers are measured in the same fashion (Figure 6.8; NRM2.14.4.1).

Where the length of a projection or an isolated pier exceeds four times the wall thickness, it is defined by NRM2 as a wall and measured accordingly (Figure 6.9; NRM2 14.5 notes, comments and glossary).

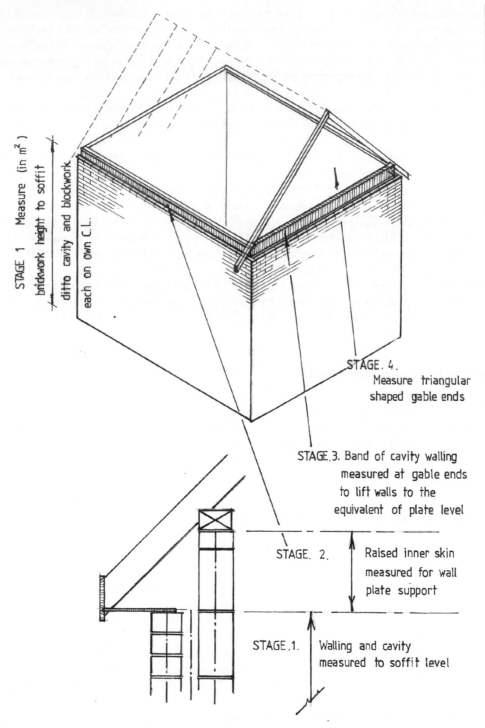

STAGE 1 Measure (in m²)
brickwork height to soffit
ditto cavity and blockwork
each on own C.L.

STAGE. 4.
Measure triangular
shaped gable ends

STAGE.3. Band of cavity walling
measured at gable ends
to lift walls to the
equivalent of plate level

STAGE. 2. Raised inner skin
measured for wall
plate support

STAGE.1. Walling and cavity
measured to soffit level

Figure 6.6 A suggest approach (showing three stages) when measuring a traditional brick and block cavity
wall for a gable ended property.

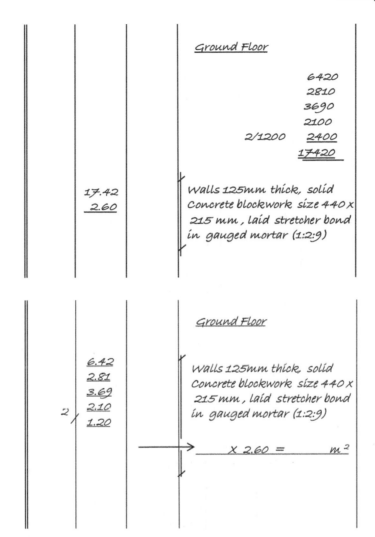

Figure 6.7 Alternative approaches to waste calculations and the presentation of booked when measuring internal partition block walling.

6.03.03.02 Arches

Arches are measured on the mean girth of the face of the arch in linear metres, stating the height of the face (NRM2 14.6). The number of identical arches, together with the width of the exposed soffit and shape of the arch (e.g. segmental/flat), must also be included in the description (Figures 6.10 and 6.11).

6.03.03.03 Closing cavities

These are most likely to occur where windows and external doors are measured, and in the normal course of events these would be included as part of the window or door opening

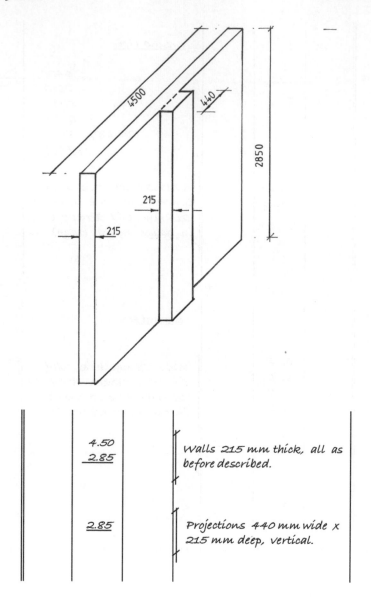

Figure 6.8 An example of an attached pier which is designated as a projection: i.e. the length of the pier is less than four times its width and is therefore measured as a projection. Hence booked as walling followed by measurement of the pier.

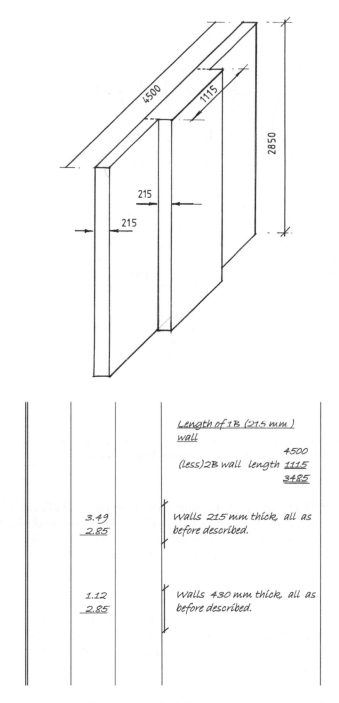

Figure 6.9 An example of what looks like an attached pier but is designated as a wall: i.e. the length of the pier is less than four times its width and is therefore defined (NRM2) as a wall. Hence it is measured as walling with two separate descriptions and two different sets of dimensions each reflecting the specific wall width.

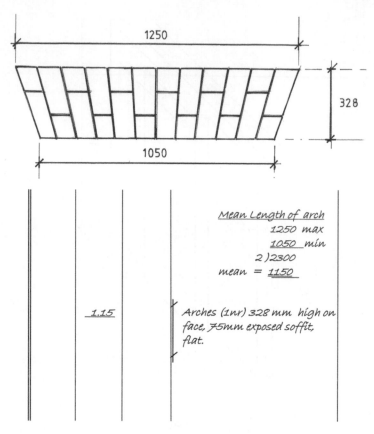

Figure 6.10 Booked dimensions and waste calculation for the measurement of a flat arch. Waste calculation necessary to establish the average length of arch.

adjustment (NRM2 14.11 and 14.12). Closing cavities are booked in linear metres as 'extra-over' the initial walling measurement, stating the width of the cavity and the method of closing together with the plane (vertical or horizontal). Any additional wall ties, insulation and damp proof course are all deemed included (Figures 6.12a and 6.12b; NRM2 14.11/12.1.2).

6.03.03.04 Ornamental bands

All are measured in linear metres stating whether they are flush, sunk or projecting (NRM2 14.7 6.13). The description should also give the width of 'set back' or 'set forward' and the plane. Bands should be described as vertical, raking, horizontal or curved (stating the mean radius on face) (see Figure 6.13).

6.03.03.05 Quoins

These are measured 'extra-over' the initial walling measurement, giving a dimensioned description together with the method of forming (NRM2 14.11, and see also NRM2 14.25;

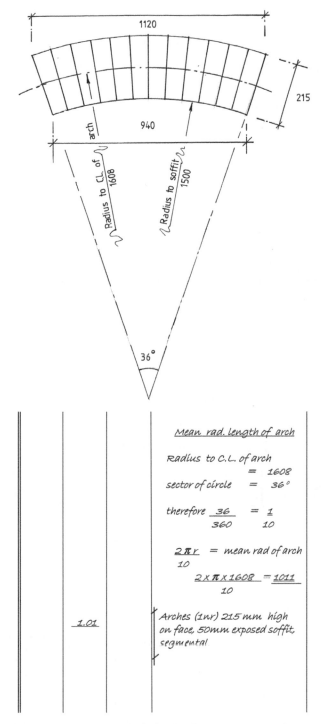

Figure 6.11 Booked dimensions and waste calculation for the measurement of a segmental arch. Waste calculation necessary to establish the average length of arch.

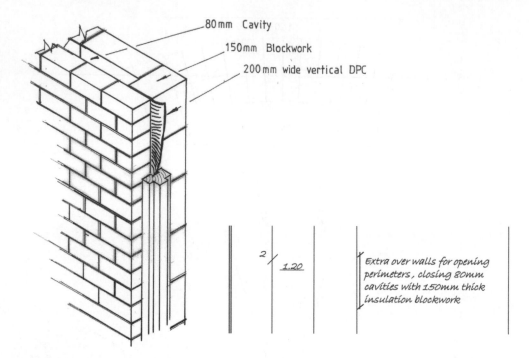

Figure 6.12a Closing cavities (blockwork).

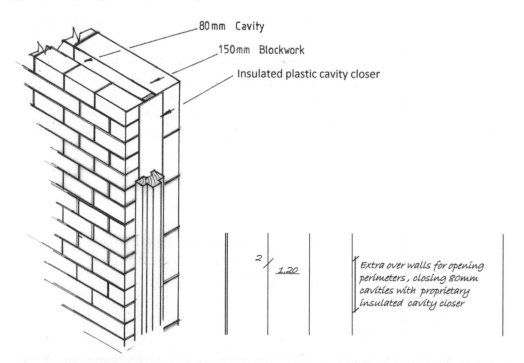

Figure 6.12b Closing cavities (proprietary cavity closer).

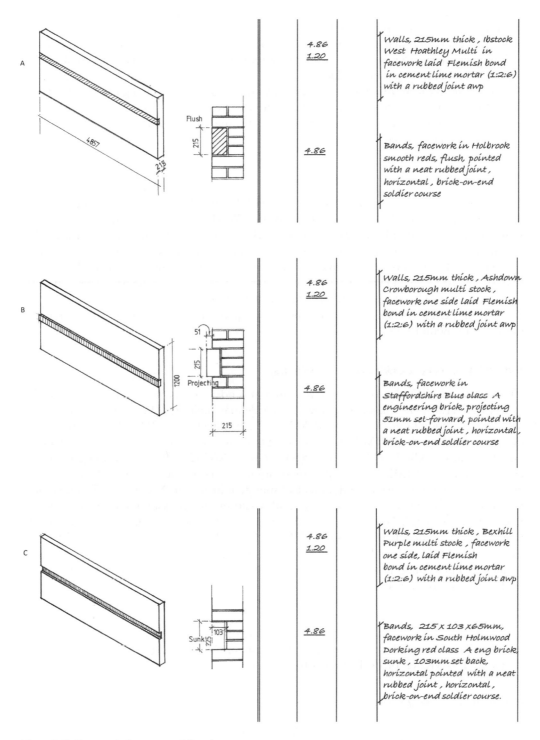

Figure 6.13 Facework Ornamental Bands.

Figure 6.14). Note that this item would only be invoked where the quoin brickwork differs from the facing bricks used in the body of the walling.

An identical approach to the measurement of quoins (i.e. measure 'extra-over' the initial walling measurement) is adopted for the measurement of forming sills, jambs, reveals, cavity closers, thresholds, steps and the like.

On occasions the architect (or the client) may not have selected a facing brick at the tender stage. In such circumstances a Prime Cost sum may be included to cover the cost of the supply of the facing bricks; since these are purchased in units of thousands, the price is usually expressed in the same fashion (Figure 6.15).

6.03.03.06 Damp proof courses

These are measured in accordance with NRM2 14.16–18. The unit of measurement is determined by the width of the damp proof course. Work up to (but not exceeding) 300 mm in width is measured in linear metres, while work that is over 300 mm wide is measured in square metres. Preformed cavity trays are measured in linear metres. In each case the gauge of the damp proof course, the number of layers and the composition and mix of the bedding materials must be given in the description together with the plane of the work (stepped work, vertical, horizontal, raking and where curved; stating the radius). Damp proof courses are deemed to include forming laps, ends and angles, pointing exposed edges and bonding to damp proof membranes.

6.03.04 Masonry work (associated items)

This section of NRM2 covers the more common sundry items associated with the work of the bricklayer. In the majority of low-rise residential buildings this would include work associated with joint reinforcement (NRM2 14.19), forming fillets (NRM2 14.20) forming and filling expansion joints (NRM2 14.22), pointing to flashings together with pointing to window and door frames (NRM2 14.21), wedging and pinning (NRM2 14.23), and forming and laying creasing courses (NRM2 14.24). All of these are measured in linear metres. Finally there is an all-embracing proprietary and individual spot item inclusion (NRM2 14.25). The unit of measurement for this is enumeration (nr) and would include measurement for such as steel lintels, steps, air bricks and weep-holes (See Figure 6.19a). A list of other items that might typically be included here is given in NRM2 14.25. It is made clear that the list is not exhaustive and that, should the need arise, any other associated masonry sundry items can be included providing these are accompanied by a dimensioned diagram and/or by stating trade brochure references.

6.03.04.01 Forming cavities, insulation

For examples of recording dimensions and descriptions for forming cavities and cavity wall insulation see NRM2 14.14 (Figures 6.16 and 6.17).

6.03.04.02 Damp proof courses

For an example of recording dimensions and descriptions for damp proof courses see NRM2 14.16–18 (Figure 6.18).

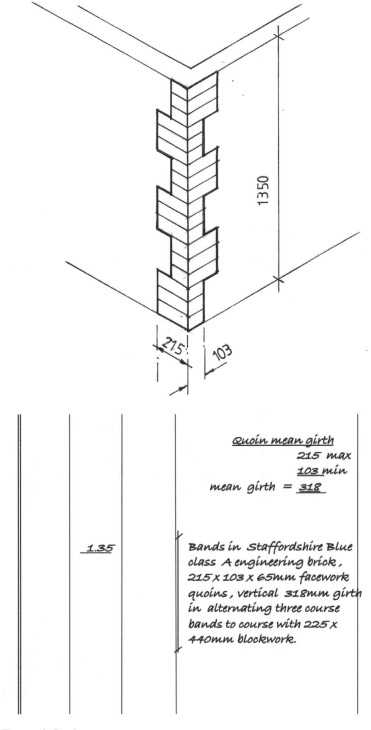

Figure 6.14 Facework Quoins.

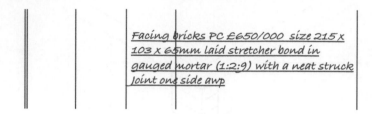

Facing bricks PC £650/000 size 215 x 103 x 65mm laid stretcher bond in gauged mortar (1:2:9) with a neat struck joint one side awp

Figure 6.15 Prime Cost sum inclusion.

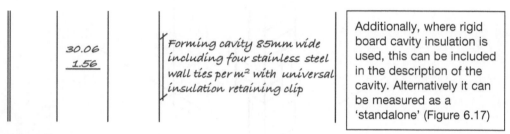

30.06
1.56

Forming cavity 85mm wide including four stainless steel wall ties per m² with universal insulation retaining clip

Additionally, where rigid board cavity insulation is used, this can be included in the description of the cavity. Alternatively it can be measured as a 'standalone' (Figure 6.17)

Figure 6.16 Forming cavity.

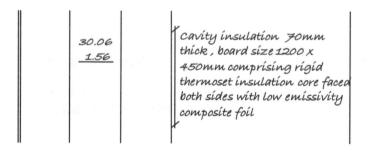

30.06
1.56

Cavity insulation 70mm thick, board size 1200 x 450mm comprising rigid thermoset insulation core faced both sides with low emissivity composite foil

Figure 6.17 Cavity wall insulation.

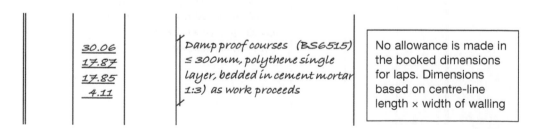

30.06
17.87
17.85
4.11

Damp proof courses (BS6515) ≤ 300mm, polythene single layer, bedded in cement mortar 1:3) as work proceeds

No allowance is made in the booked dimensions for laps. Dimensions based on centre-line length × width of walling

Figure 6.18 Damp proof course.

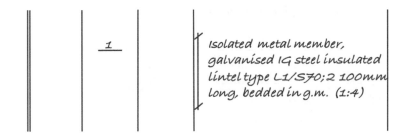

Figure 6.19a Metal lintel.

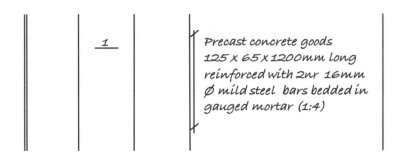

Figure 6.19b Precast concrete lintel.

6.03.04.03 Proprietary items (metal lintels)

Metal lintels in external walls are most likely to be associated with window and door openings, and are measured as part of the window or door 'opening adjustment' (see Chapter 11). Although lintel lengths (and possibly heights) will vary throughout a building, a specific lintel manufacturer will be identified. Reference should be made to the proprietary manufacturer's catalogue, giving product-specific references in the descriptions as appropriate. Where a number of similar components are involved (as with steel lintels), it is advisable to set up a heading on dimension paper recording generic details. This eliminates the need to repeat the full detail of the component information with each description. Instead, the measurer need only record the variable details. For steel lintels this would be the lintel length and manufacturer's code.

See NRM2 14.25 (Figure 6.19a); for precast concrete lintels see NRM2 13.1 (Figure 6.19b).

6.04 Two-storey detached residential property with attached garage

The measurement example that follows is based on a traditional-build, two-storey detached residential property. From a measurement perspective, the attached single-storey pitched roof garage provides an opportunity to use geometric formulae and demonstrate 'an adjustment' for the change of walling specification at the junction between the living accommodation and the garage.

6.04.01 Plan, section, specification notes

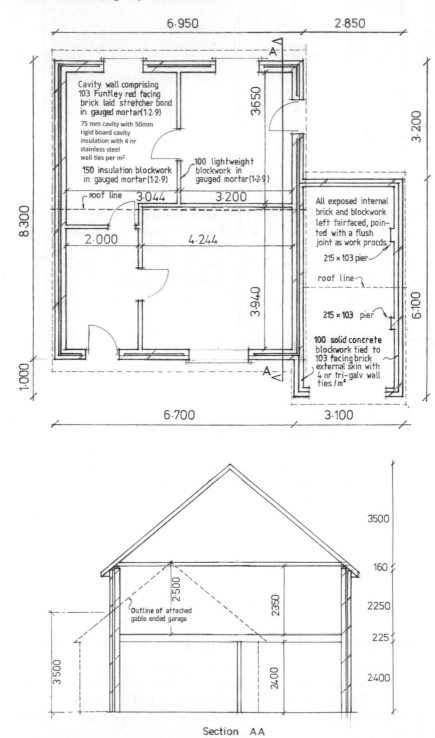

Section AA

6.04.02 Measurement example Job ref LJAS 17

Masonry – Detached Two Storey Dwelling with Attached Garage; Job ref: LJAS 17; Page 01 of 06

Take off list

Cavity Walls
- External facing brickwork
- Forming cavity
- Insulation
- Internal insulation blockwk

Composite (garage) Walls
- External facing brickwork
- Solid block internal wall

- Adjustment of facework for garage
- Projections
- Internal ground floor partitions

Centre-line of ext brick wall

External girth brickwork

	2/6950	13 900
	2/8300	16 600
Ext girth bwk =		30 500
CL of ext brick wall		
Less 4/2/ $\frac{1}{2}$/103		412
CL of ext skin =		30 088

Brickwork height to soffit

Dpc to first floor	2 400
floor joists	225
Joists to soffit	2 250
Bkw ht to soffit	4 875

2/30.09
2/$\frac{1}{2}$/4.88
2/8.30
0.16
8.30
3.50

Walls 103 mm thick in
Funtley red facing bricks
laid stretched bond in
gauged mortar (1:2:9)
pointed with a neat rubbed
joint as work proceeds in
skins of hollow walls.

{ Gable raising
{ Gable end

Measurement taken over all openings; adjustment for windows and door openings made later (ref Chapter 11 measurement of Joinery /Windows and Doors)

Initial measurement for two-storey dwelling; attached garage block measured separately later.

PLAN

DWELLING

ATT. GAR.

NRM2 14.1.1.1

Masonry – Detached Two Storey Dwelling with Attached Garage; Job ref: LJAS 17:   Page 02 of 06

Centre line of cavity

Ext girth bwk   =   30 500

	103	Less	
75	37	4/2/140	
2	140		1 120

CL of Cavity   29380

> The same dimensions have been
> adopted for insulation and cavity.
> Note that it is also possible to measure
> each separately on an individual centre line

NRM2 14.14.1.1

	29.38	Forming cavities
2/	4.88	75 mm wide inc.
	8.30	4nr. stainless steel
2/1	0.16	wall ties /m²
2/	8.30	
	3.50	

(Gable
  raising

(Gable
   end

&

Rigid board cavity wall
insulation 50mm thick,
fixed with wall clips,
joints
taped.

NRM2 14.15.1.1

Centre line of internal block
wall
Ext girth bwk   =   30 500

	103
	75
150	75
2	228

Less   4/2/228   2 304
CL of int blockwk   28 196

Blockwork height to wall plate
Dpc to first floor   2 400
    floor joists   225
ff Joists to clg   2 350
Bkw ht to soffit   4 975

Less plate 50
       bed   10   60
                      4 915

<u>Blockwork raising length</u>

External length of facings
8 300

103
75
178   2/178
356
7944

		Walls 150mm thick insulation blockwork laid stretcher bond in gauged mortar (1:2:9) with a raked joint in skins of hollow wall	Gable raising / Gable end
	28.20		
2/	4.92		
	7.95		
2/1	0.16		
2/	7.95		
	3.50		

NRM2 14.1.2.1

<u>Attached Garage</u>
External brick wall girth
2 850
6 100
3 100
1 000
13 050
Less 3/103   309
Garage ext brickwk CL 12.741

Garage projection return gable height
span x tangent 40°
= 1.000 x 0.839
= 0.839

		Walls 103 mm thick in Funtley red facing bricks laid stretched bond in gauged mortar (1:2:9) pointed with a neat rubbed joint as work proceeds built against other work including 4nr stainless steel wall ties to course with blockwork	
	12.74		
1/	2.40		
2/	6.10		
	2.50		
	1.00		
	0.84		

PART PLAN — DWELLING / ATT. GARAGE
1 CL adjusted by only external angles
3   2

PART ELEVATION SHOWING GARAGE PROJECTION

Return part gable to garage
0.83
1.00

NRM2 14.1.1.4

Masonry – Detached Two Storey Dwelling with Attached Garage; Job ref: UAS 17; Page 04 of 06

Attached garage (Contd)

Centre –line blockwork
    external girth (as before)
                  13 050

Less   103
150   } 75
 2   178
       3/2/178 1 068
                 11 982

Gable raising blockwork
length
               6 100
   less 2/103   206
               5 894

Return gable blockwork height
        span x tan 40°
  (1000 – 103) x 0.839
       = 0.753

2/1	11.98	Walls 100 mm thick solid
2/	2.40	concrete blockwork laid
	5.89	stretched bond in gauged
	2.50	mortar (1:2:9) pointed
	0.90	with a neat rubbed joint as
	0.75	work proceeds built
		against other work (wall
		ties included elsewhere)

Adjustment for blockwork in
lieu of facing brickwork to
internal garage walling

   Height to garage ridge
           2 400
             225
           2 625
           5 250
Garage length A
    6 100 = 3 050
      2
Garage length B   6 100
    3 050
less   1 000   4 050
             2 050

NRM2 14.1.2.4

Section through garage

<u>Adjustment for blockwork in lieu of facing brickwork to internal garage walling (Contd)</u>

Average Height A	5 250
	2 400
	2) 7 650
→	= 3 825
Average Height B	5 250
	3 500
	2) 8 750
→	= 4 375

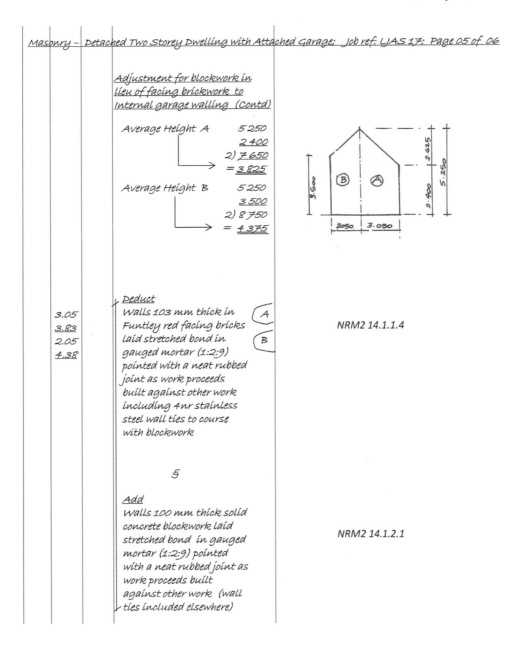

<u>Deduct</u>

3.05
3.83
2.05
4.38

Walls 103 mm thick in Funtley red facing bricks laid stretched bond in gauged mortar (1:2:9) pointed with a neat rubbed joint as work proceeds built against other work including 4nr stainless steel wall ties to course with blockwork (A) (B)

NRM2 14.1.1.4

&

<u>Add</u>

Walls 100 mm thick solid concrete blockwork laid stretched bond in gauged mortar (1:2:9) pointed with a neat rubbed joint as work proceeds built against other work (wall ties included elsewhere)

NRM2 14.1.2.1

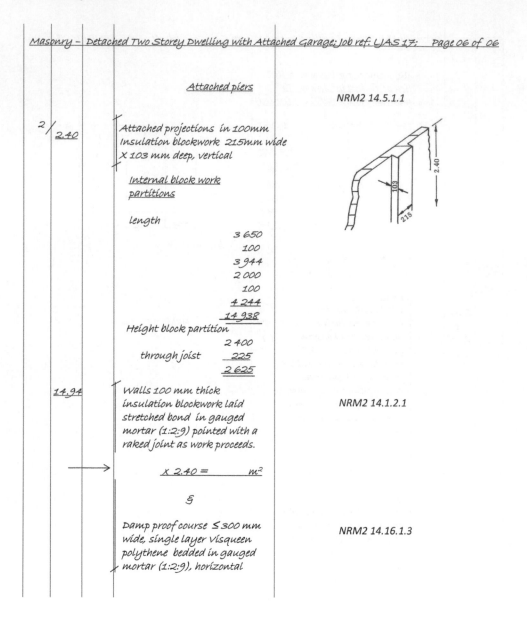

Attached piers

NRM2 14.5.1.1

2/ 2.40

Attached projections in 100mm
Insulation blockwork 215mm wide
X 103 mm deep, vertical

Internal block work
partitions

length

	3 650
	100
	3 944
	2 000
	100
	4 244
	14 938

Height block partition

	2 400
through joist	225
	2 625

14.94

Walls 100 mm thick
insulation blockwork laid
stretched bond in gauged
mortar (1:2:9) pointed with a
raked joint as work proceeds.

NRM2 14.1.2.1

X 2.40 =   m²

&

Damp proof course ≤ 300 mm
wide, single layer visqueen
polythene bedded in gauged
mortar (1:2:9), horizontal

NRM2 14.16.1.3

7 Concrete-framed buildings

7.01 Introduction to framed buildings

This chapter and the next (Chapter 8) will review the techniques and approaches adopted when using NRM2 for the measurement of concrete and steel-framed buildings. While this is a departure from the residential 'theme' that runs through the other chapters of this text, the techniques adopted will be familiar. Depending on the construction method in question, any one of three (or more) NRM2 work sections may be deployed when measuring a concrete-framed building, including NRM2 11 In situ Concrete, NRM2 12 Precast/ Composite concrete and NRM2 13 Precast Concrete. For the measurement of a steel-framed building, NRM2 work section 15 Structural Metalwork gives the recommended procedure for recording dimensions (see Chapter 8).

7.02 In situ concrete-framed buildings

NRM2 Work Section 11 (In situ Concrete) also includes the measurement rules for the other necessary main components of an in situ concrete-framed building such as reinforcement (bar and mesh) and temporary support or formwork. As with other work sections, there is a requirement that the measured items are accompanied by the following minimum information:

general arrangement drawings; the relative position of all members together with their respective sizes; the thicknesses of all slabs; and the permissible loads in relation to casting times. All of these details should be available from both the structural engineer's drawings and accompanying specification details. The mandatory information that needs to be provided includes the kind and quality of materials, details of testing of materials/finished work, any limitations on method, sequence, speed or size of concrete pouring. In addition and as and where appropriate, the method of compacting, the method of curing and any details of achieving watertightness should also be given in the description.

7.03 Reinforced in situ concrete

NRM2 provides a single set of rules for the measurement of all in situ concrete work. This includes plain in situ concrete, reinforced in situ concrete, fibre-reinforced in situ concrete and sprayed in situ concrete. The principal unit of measurement in each case is cubic metres. In previous editions of the Standard Method of Measurement, the classification of 'types of concrete work' in framed buildings was based on components – i.e. in slabs, in beams, in columns, etc. NRM2 adopts a slightly different approach and classes these as: horizontal work, vertical work, sloping work ≤15°, or sloping work ≥15°. Even so, there is an option that allows for horizontal work to be aggregated as a single volume or presented as separate components (NRM2 11.2.*.*.*.*.1 and 2). A full description would also need to include a reference to the overall thickness of the work (≤300 mm thick, ≥300 mm thick) and whether the work was in structures or in blinding (Figure 7.1).

For sloping work, a further classification of 'in staircases' is available. Where horizontal or sloping work is poured against earth or unblinded hardcore, or where the volume of reinforcement exceeds 5% by volume of the overall, this will need to be described (NRM2 11.2/3/4.*.*.1/2). The following (Figures 7.2, 7.3 and 7.4) are examples of typical reinforced in situ concrete-framed building descriptions.

The measurer faced with the prospect of booking dimensions for a reinforced in situ concrete-framed building should first take time to study the plans, sections and elevations so that they have a clear idea in their mind as to how the structure will be constructed. Typically the concrete frame of a building would be based on a repeating 'grid layout' with a regular set of plan dimensions that run from the centre of each vertical column. In addition, when considering vertical sections and elevations, there is likely to be a standard floor-to-ceiling (or floor slab to floor slab) height. This provides a repeating pattern that will allow the measurer to identify a systematic and logical approach to recording dimensions. Approaches may differ based on individual preference and/or office practice. At its simplest, a framed building can be broken down into three components: horizontal floor slabs with attached beams and vertical columns. An example of booking dimensions for a reinforced in situ concrete frame is provided at the end of this chapter, and has adopted the following approach when recording dimensions for in situ concrete work in structural frames.

- Book dimensions for the in situ concrete horizontal slab (plan area dimensions × slab thickness to give volume) initially for one floor. Use the 'timesing' column to multiply this volume by the respective number of floors.
- Book dimensions for any attached beams (both perimeter and intermediary) associated with the same single-floor slab (also a volume). Do this for one floor and then use the timesing column to pick up the beams associated with each 'repeating' floor. Remember

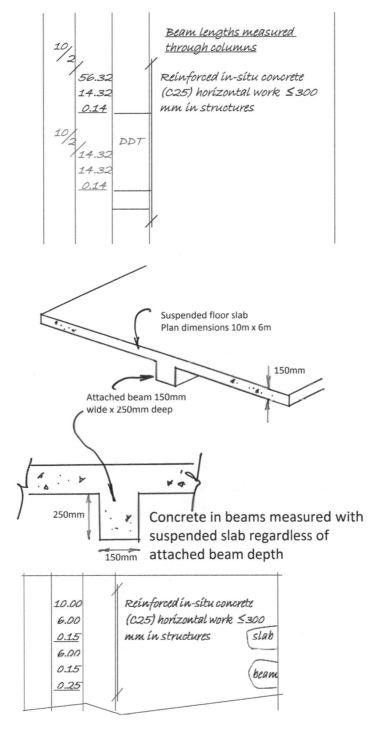

Figure 7.1 Measuring slabs and attached beams.

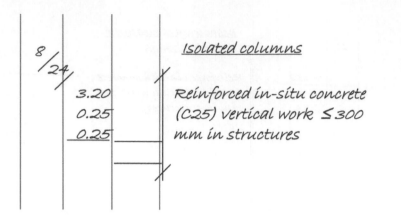

Figure 7.2 Typical descriptions reinforced concrete isolated columns.

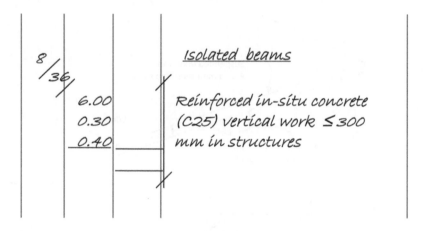

Figure 7.3 Typical descriptions reinforced concrete isolated beams.

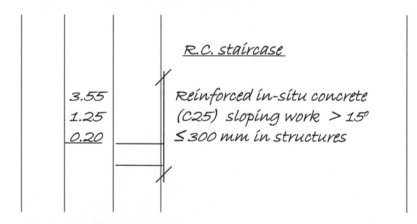

Figure 7.4 Typical descriptions reinforced concrete staircase.

that the 'grid' pattern of these beams will mean that they intersect each other. At this stage we have measured the beams without considering any overlaps and made no allowance for the point where they intersect. An adjustment (deduction) will need to be made to accommodate this. As previously, if we make the adjustment for one floor we need to use the timesing column to multiply this by the respective number of floors.

• So far we have only measured the horizontal members (slabs and attached beams). The columns (vertical members) can now be measured. These are booked as a volume based on the column's plan dimensions and the vertical (floor-to-ceiling) height. Once again, we only need to consider a single floor. We also have to remember that we have already measured the floor slab and attached beams, so a waste calculation to establish the vertical height of the column need only be from the top of the floor slab to the underside of the attached beams associated with the storey immediately above. Once we have recorded the dimensions for the columns associated with a single floor, we can use the timesing column to multiply this out by the respective number of storeys.

7.04 Formwork (shuttering)

In situ concrete, unlike precast concrete, requires temporary support while the wet concrete mixture 'sets' or cures. Traditionally the 'moulds' that provide this temporary support have been made of shuttering-grade plywood with sawn timber to give rigidity to the formwork and 'acrow-props' to transfer the weight of both the in situ concrete and the temporary support to a load-bearing plane. Increasingly lightweight modular formwork systems made from high-tensile steel, aluminium, fibreglass or certain types of plastic make for a more efficient and cost-effective approach.

In regard to the measurement of formwork, the materials/techniques adopted to provide temporary support are considered a 'contractor's risk'. As such, the measurer need not be concerned with the approach adopted by the contractor and simply needs to record dimensions (in accordance with NRM2) for the various edges, surfaces and soffits of the finished structure. The one exception to this would be where a specific system of temporary support is specified by the structural engineer/architect.

There are twenty different classifications for the measurement of formwork identified in NRM2 (NRM2 11.13–11.32), with three different units of measurement adopted: metres, square metres and enumerated items. By way of an introduction to the techniques and approaches related to the measurement of formwork, this chapter will consider examples associated with a typical multi-storey, in situ concrete-framed building. It is assumed that any formwork is grouped with the same set of mandatory information required by NRM2 for the measurement of in situ concrete work. In addition, attention should be paid to the notes, comments and glossary of NRM2 works section 11 (formwork), which (by way of example) includes the following.

• Where a 'plain' finish to the surface of concrete is specified, this will be left to the discretion of the contractor. So if anything other than a plain finish is required to concrete surfaces, it will require describing as 'special finish formwork'.
• Permanent formwork or formwork left-in is so described.
• No deductions are made for voids ≤5.00 m^2.
• All kickers, except to walls, are deemed included.

7.04.01 Formwork to soffits of slabs

Formwork to the soffits of an in situ concrete slab would be described as 'soffits of horizontal work', and is measured in square metres. The formwork description should also include reference to the thickness of the in situ concrete being supported in one of the following three categories:

- for concrete ≤300 mm thick;
- for concrete 300–450 mm thick;
- for concrete >450 mm thick.

Finally, the description must make reference to the propping height (floor-to-ceiling) in one of the following categories:

- propping ≤3 m high;
- propping over 3 m but not exceeding 4.5 m high;
- and thereafter in stages of 1.5 m.

7.04.02 Formwork to the sides and soffits of attached beams

An attached beam (as the title suggest) is a beam that forms part of a suspended slab. In normal circumstances this would form part of the structural integrity of the frame and would sit below the slab. Measurement of the formwork necessary to provide an attached beam is measured in square metres (NRM2 11.18). A waste calculation, based on the attached beam sectional dimensions, will be necessary to establish the 'girth' of the attached beam (this will give one of the two necessary dimensions required), and the length of the same attached beam will be evident from the plans and general layout drawings provided (Figure 7.5).

The description for any formwork associated with an attached beam (assuming a plain finish) would need to include the class of formwork (all as per NRM2 11.18) 'sides and soffits of attached beams'. This is followed by a reference to the cross-sectional shape of the beam, of either regular shape (defined as square or rectangular) or irregular shape (anything other than square or rectangular). In the last instance an irregularly shaped attached beam will need to be accompanied by a dimensioned description or diagram. The final part of the description makes reference to the propping height and adopts the same approach used when describing formwork to soffits of horizontal work (suspended slabs). This gives the following three options: propping not exceeding 3 m high, propping over 3 m but not exceeding 4.50 m high and thereafter in stages of 1.5 m.

7.04.03 Formwork to sides of isolated columns

An isolated column is a vertical support that is able to transfer the load of floor slab/s to a safe load-bearing foundation or substructure. Formwork to isolated columns adopts the same set of rules used for attached beams (see above). As with attached beams, formwork to isolated columns is measured in square metres (NRM2 11.20) and, in similar fashion, a 'girthing' waste calculation will be necessary based on the cross-sectional dimensions of the column. This will be used alongside the vertical height of the formwork as booked dimensions to generate the area of formwork required. Establishing the vertical height is likely to follow the same waste

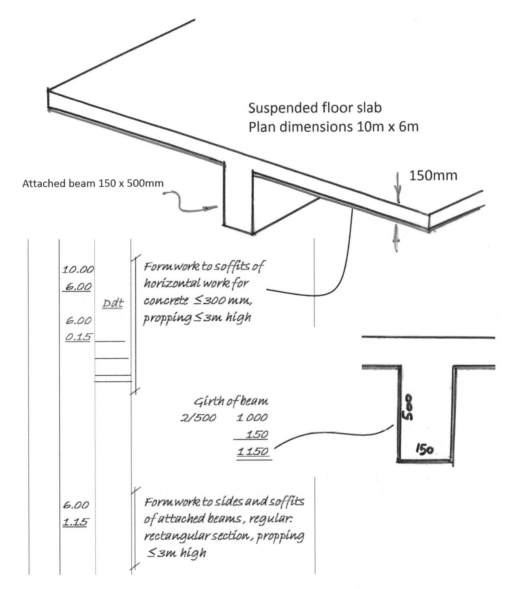

Figure 7.5 Formwork to soffits of slab and formwork to sides and soffits of beams.

calculation prepared for the vertical height of the column (from the top of the floor slab to the underside of attached beams or suspended slab).

The only difference is that the measurer is required to identify the number (nr) of columns as part of the description. In most situations, isolated column heights will be consistent throughout the frame of the building. Where there are a number of different column heights in any one structure, each distinctive 'height set' of columns would need to be measured separately in order to accommodate this distinction (See NRM2 11.20).

7.05 Reinforcement

Although concrete has a high compressive strength, it is relatively weak under tension. To counter this, steel reinforcement bars or sheets of fabric (mesh) reinforcement are carefully positioned in a floor slab, beam, column or other structural member. The type, size and positioning of reinforcement is determined by a structural engineer, who will prepare drawings and schedules detailing the exact nature of the proposed structure. For the person responsible for measuring reinforcement, the availability of a reinforcement schedule is a bonus. Armed with this, an understanding of how to 'read' a bar reinforcement schedule together with a grasp of NRM2 Work Section 11 (recording dimensions for bar reinforcement) is much simplified.

As noted previously there are two types of reinforcement – bar reinforcement and fabric/mesh reinforcement. In each case the reinforcing bar could be made from either mild or high-tensile steel. Both the drawings and the schedule will identify the different types of steel reinforcement with one of the following prefixes before the diameter of the reinforcement bar. The letter 'R' is used to denote mild steel, while 'T' or 'Y' indicate high-tensile and high-yield steel, respectively. All bar reinforcement is manufactured to standard diameters (6, 8, 10, 12, 16, 20, 25, 32 and 40 mm).

In the case of fabric/mesh reinforcement, these are produced in standard sheets with the reinforcing bars welded at 90 degrees to each other to form an 'open grid'. The unit for measuring bar reinforcement is tonnes (t), and for fabric or mesh reinforcement is square metres (m^2).

7.05.01 Bar reinforcement

The first record (or booking) of dimensions when measuring bar reinforcement will be in linear metres. These initial lengths of steel bar are then converted from metres to tonnes using standard steel bar weight tables. These work on the principle that a specific diameter of steel bar has a known weight for every metre of bar length (expressed in kg/m). To enable conversion (and in order to accord with NRM2 11.33.1/34.1), the varying diameter of steel bars will need to be booked as separate entries on dimension paper. See bar reinforcement weight conversion (Figure 7.6) and sample bar reinforcement schedule (Figure 7.7)

In addition to a bar reinforcement schedule, the structural engineer may include an annotation on a general layout drawing expressed as a set of reference numbers and letters. Typically this will appear as follows (see Figure 7.8 and accompanying annotation for explanation of coding).

All reinforcement requires a minimum concrete cover to exposed edges (exposure causes deterioration of the metal bar). This has to be factored into any waste calculations used to establish bar length. The amount of concrete cover will vary depending on the aggregate size of the concrete and the degree of exposure, although a minimum of 25 mm would usually be recommended for non-exposed locations and 50 mm for exposed locations. Reference to either the drawing or the specification will identify the minimum concrete cover required for reinforcement. In its simplest form, this would mean that a straight bar in a suspended concrete slab of 4.800 m with 50 mm of concrete cover specified would require a reinforcing bar length of 4.800 mm less twice times 50 mm (i.e. 4.800 less (2/50) = 4.700 m). It should also be noted that reinforcement bars are manufactured in a number of standard shapes referred to as 'bar shapes' in accordance with BS 8666/EN10080. Even though reinforcement bars may be described and referred to as 'straight', this definition also denotes bar shape codes that are straight bars with bent ends (shape codes 11 and 12) or hooked ends (shape code 13). Many other standard manufactured shape codes are available, and these give the overall length of bar used as a function of the main dimensions (see latest BS and EN standards). NRM2 requires reinforcing bars to be described in one of the following classes: straight, bent, curved or links. Tying wire, chairs, spacers and forming hooks are deemed included. In addition, bars exceeding 12 m in length, deformed bars and any bending restrictions will also need to be included in the description. Wherever possible it is recommended that booked dimensions are based on bar reinforcement schedules and/or standard bar shape codes.

bar diam	weight kg/m
6 mm	0.222
8mm	0.395
10mm	0.616
12mm	0.888
16mm	1.579
20mm	2.466
25mm	3.854
32mm	6.313
40mm	9.864
50mm	15.413

Figure 7.6 Bar reinforcement weight conversion table.

SJL & Partners   Job nos: 170487
Project: Staplewood Centre   drwg ref: ADP   Last Rev: 04 11 25
Client: DMWSL 613 Limited   date: 30 06 25

Member	bar mark	type and size	nos of mbrs	nos of bars in each	total nos	length of each bar + mm	A* mm	B* mm	C* mm	D*mm	EorR* mm	Rev letter
column 3	5	H16	450	4	1,800	3760						
atchd bm 2	6	H8	10	654	6,540	4 800						

location refences nos — unique bar refernce — H = high Yield Steel + bar diam — nos of members — nos of bars in each member — total nos of bars — length of bar

Figure 7.7 Example bar reinforcement schedule.

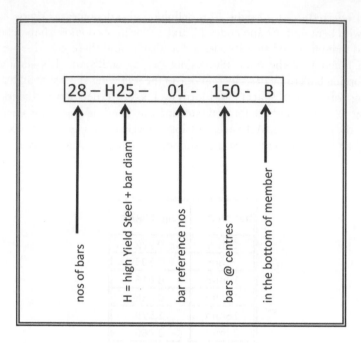

Figure 7.8 Referencing for bar reinforcement.

7.05.02 *Fabric or mesh reinforcement*

Sometimes referred to as welded fabric, this is a factory-prepared sheet of bars delivered to site in 2.40 × 4.80 m sheets. Alternatively it can be prepared in sheets cut to a specific size or shape. Typically this would be specified for a structural component like an in situ concrete floor slab and would be booked in square metres (NRM2 11.37). It is often referred to by an alpha/numeric code (e.g. A252, B385), where the letters A, B, C or D identify the mesh aperture size/shape and the numbers give the bar thickness (or more precisely the longitudinal bar area expressed in mm^2/m). The description should give the weight of the sheet of mesh reinforcement (expressed in kg/m^2), the fabric reference code and the minimum extent of any side and end laps. Any bent fabric reinforcement (e.g. any wrapping to structural steel) and any strips of fabric in one width need to be so described (NRM2 1137.*.1/2). The following items are deemed included: laps, tying wire, all cutting and bending. Accessories comprising spacers, stools, chairs and other supports are also deemed included. Any voids of less than 1.00 m^2 are ignored when booking dimensions (NRM2 11.37.*.*.*.2).

7.06 Ten-storey concrete-framed building

7.06.01 Plan, section, specification notes

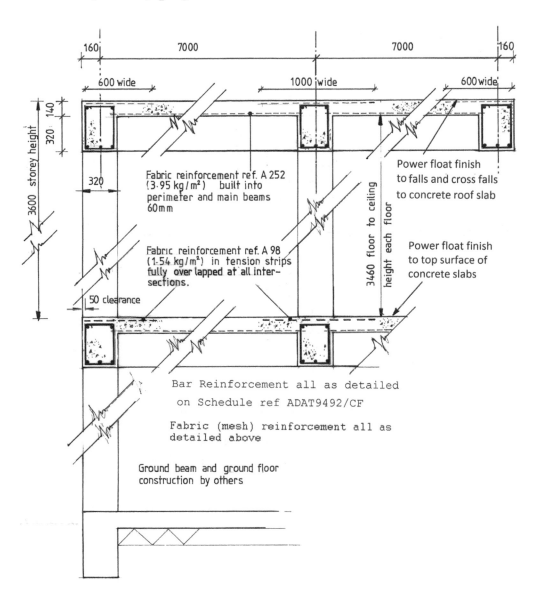

160   7000   7000   160

600 wide   1000 wide   600 wide

140   320

3600 storey height

320

Fabric reinforcement ref. A 252 (3·95 kg/m²) built into perimeter and main beams 60mm

Fabric reinforcement ref. A 98 (1·54 kg/m²) in tension strips fully over lapped at all inter-sections.

50 clearance

3460 floor to ceiling height each floor

Power float finish to falls and cross falls to concrete roof slab

Power float finish to top surface of concrete slabs

Bar Reinforcement all as detailed on Schedule ref ADAT9492/CF

Fabric (mesh) reinforcement all as detailed above

Ground beam and ground floor construction by others

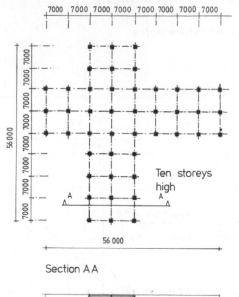

Section A A

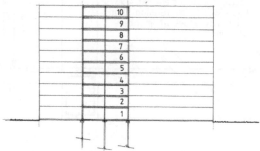

Bar Reinforcement all as detailed on Schedule ref ADAT9492/CF

Reinforcement Schedule ADAT9492/CF

SJL & Partners				Job nos: ADAT 9492/CF						
Project: Staplewood Centre				drwg ref: ADP Last Rev: 300628						
Client: DMWSL 613 Limited				date: 300625						

Member	bar mark	type and size	nos of mbrs	nos of bars in each	total nos	length of each bar + mm	A* mm	B* mm	C* mm	D*mm	EorR* mm	Rev letter
Slab (L)	01	R12	10	370	3700	6000						
Slab (W)	02	R12	10	286	2860	7540						
Beam	03	R16	150	5	750	8120						
Column	04	R16	450	4	1800	3600						
Stirrups	05	R10	150	4	600	1040						
Links	06	R10	450	18	8100	1560						

7.06.02 Measurement example Job ref: ADAT 9492/CF

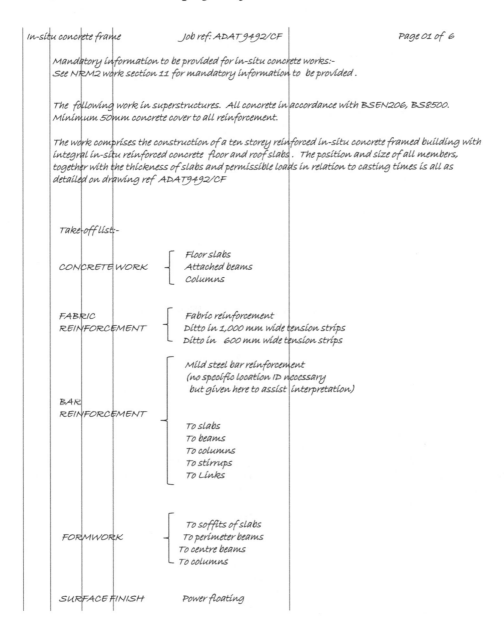

In-situ concrete frame   Job ref: ADAT 9492/CF   Page 01 of 6

Mandatory information to be provided for in-situ concrete works:-
See NRM2 work section 11 for mandatory information to be provided.

The following work in superstructures. All concrete in accordance with BSEN206, BS8500.
Minimum 50mm concrete cover to all reinforcement.

The work comprises the construction of a ten storey reinforced in-situ concrete framed building with
integral in-situ reinforced concrete floor and roof slabs. The position and size of all members,
together with the thickness of slabs and permissible loads in relation to casting times is all as
detailed on drawing ref ADAT9492/CF

Take-off list:-

CONCRETE WORK
- Floor slabs
- Attached beams
- Columns

FABRIC REINFORCEMENT
- Fabric reinforcement
- Ditto in 1,000 mm wide tension strips
- Ditto in 600 mm wide tension strips

BAR REINFORCEMENT
Mild steel bar reinforcement
(no specific location ID necessary
 but given here to assist interpretation)
- To slabs
- To beams
- To columns
- To stirrups
- To Links

FORMWORK
- To soffits of slabs
- To perimeter beams
- To centre beams
- To columns

SURFACE FINISH   Power floating

In-situ concrete frame　　　　*Job ref: ADAT 9492/CF*　　　　　　*Page 02 of 6*

Overall slab dimensions

Width/length　　　　　　56 000

Add
Perim beam 2/160　　　　320
　　　　　　　　　　　　56 320

Breadth　2/7000　　14 000
Add
perim beam　2/160　　　320
　　　　　　　　　　　　14 320

Slab thickness
　storey height　　　　　3 600
　floor to ceiling height　3 460
　　slab thickness　　　　140

Beam lengths (measured through columns)

10/2/	56.32		Reinforced in-situ concrete (C25) horizontal work ≤300 mm in structures	NRM2 11.2.1.2
	14.32			
	0.14			
10/2/	14.32	DDT		
	14.32			
	0.14			

Beam lengths
(measured through columns)

10/6/	56.32		Reinforced in-situ concrete (C25) horizontal work ≤300 mm in structures	NRM2 11.2.1.2
10/2/6/	0.32			Long beam
	0.32			Short beam
	14.32			
10/2/6/	0.32			
	0.32			short beam o'lap with long
10/2/6/3/	0.32	DDT		
	0.32			Long beam o'lap with long
10/3/3/	0.32			
	0.32			
	0.32			
	0.32			

In-situ concrete frame   Job ref: ADAT 9492/CF   Page 03 of 6

<u>Column Heights</u>

	3 600

<u>Less</u>
Slab 140
Beam <u>320</u>   <u>460</u>
 Col ht to u'side of beam <u>3 140</u>

<u>Number of columns per floor</u>

9/3	27
6/3	<u>18</u>
	<u>45</u>

10/45		Reinforced in-situ concrete (C25)	NRM2 11.5.2.1
	3.14	Vertical work > 300mm thick in	
	0.32	structures	
	<u>0.32</u>		

<u>Bar Reinforcement</u>
Information based on StEng's bar
reinforcement schedule
The following mild steel bar
reinforcement (BS4449)

Bar reinforcement measured in linear metres here and converted to tonnes once transferred to Abstract

3700	6.00	12mm Ø straight	slab L 01	NRM2 11.33.1.1
2860	7.54		slab W 02	
750	8.12	16mm Ø straight	beam	NRM2 11.33.1.1
1800	3.60		column	
600	1.04	10 mm Ø straight	stirrups	NRM2 11.33.1.4
8100	1.56		links	

Alternatively can be converted to tonnes on dimension page by applying respective kg weight of bar per metre

<u>Fabric Reinforcement</u>

In bottoms of slabs (A252)
     each bay @   7 000
        Less 2/160   <u>320</u>
                         6 680
     add o'lap 2/60   <u>120</u>
                         <u>6 800</u>

Note: 28 floor slab bays per floor

| In-situ concrete frame | Job ref: ADAT 9492/CF | Page 04 of 6 |

Fabric Reinforcement (Contd)

10/28/ 6.80
         6.80

Mesh reinforcement BS4483 ref A252 weighing 3.95kg/m2 minimum overlap of 200mm (one mesh square)

NRM2 11.37.1/2/3

Fabric Reinforcement in tension strips

Perimeter tension strip

4/56 320	225 280
Less cover	
4/2/50	400
	222 880

In this case the perimeter length of the fabric tension strip is based on the external perimeter length. Alternatively it could be based on the CL

10/222.48
    0.60

Mesh reinforcement BS4483 ref A98 weighing 1.54kg/m2 in strips in one width 600 mm wide minimum overlap of 200mm (one mesh square)

NRM2 11.37.1/2/3.2

Fabric Reinforcement to long beam and cross beam in tension strips

Beam length (max) 56 320
Less cover
       2/50   100
                56 220

Cross beams   14 320
Less cover
       2/50   100
                14 220

10/2/56.22
10/12/1.00
     14.22
      1.00

Mesh reinforcement BS4483 ref A98 weighing 1.54kg/m2 in strips in one width 1 000 mm wide minimum overlap of 200mm (one mesh square)

NRM2 11.37.1/2/3.2

In-situ concrete frame   Job ref: ADAT 9492/CF   Page 05 of 6

<u>Formwork</u>
Slab bay dimensions

		7 000
Less	2/160	320
		6 680

10/28/ 6.68
6.68

Formwork to soffits of horizontal
work for concrete ≤ 300mm
thick, propping over 3.00m but
not exceeding 4.50m high.

NRM2 11.15.1.2

Formwork to edge beam and
perim of slab

<u>Girth of beam</u>	3/320	960
	slab	140
		1 100

Part section through edge beam

Girth of formwork to beam

Edge beam perim length from
t/off page 4   225 280

Centre line
less 4/320   1 280
                    224 000

224.00
1.10

Formwork to sides and soffits of
attached beams regular shaped,
square, propping over 3.00 m not
exceeding 4.50 m high

NRM2 11.18.1.2

<u>Formwork to centre beams and
cross beams</u>

Centre beam overall length
from t/off page 2   56 320
      less 2/320   640
                       55 680

Cross beam o'all length
from t/off page 2   14 320
      less 2/320   640
                       13 680
Less centre beam   320
                       13 360

<u>Girth of centre and cross beams</u>

| | 3/320 | <u>960</u> |

10/2/	55.68
10/12/	0.96
	13.36
	0.96

Formwork to sides and soffits of attached beams regular shaped, square, propping over 3.00 m not exceeding 4.50 m high   (Centre bms)   (Cross bms)

NRM2 11.18.1.2

<u>Formwork to columns</u>

<u>Column height</u>
    from t/off p3   <u>3140</u>

<u>Column girth</u>   4/320   <u>1280</u>

<u>Number of columns per floor</u>

3 columns/grid line
        length   3/9   27
        width   3/9   <u>27</u>
                        54
<u>less</u> o'count @ central bay
                3/3   <u>9</u>
Number of columns/floor   <u>45</u>

| 10/45/ | 3.14 |
| | 1.28 |

Formwork to sides of isolated columns (450 nr) regular shaped, square, propping over 3.00m but not exceeding 4.50m high.

NRM2 11.20.1.2

| 10/ | 56.32 |
| | 14.32 |

Power floating to top surfaces   (floors 1 - 10)

NRM2 11.9.1

| 56.32 |
| 14.32 |

Power floating to top surfaces to falls and cross falls   (roof)

NRM2 11.9.2/3

8 Steel-framed buildings

8.01 Introduction

When compared to concrete framed structures, a steel frame building would appear to have many advantages. It is lighter, stronger and facilitates a speedier erection. In addition, the comparatively light weight of a steel-framed structure allows for savings in foundation costs. However, there is a problem with steel in that although it is not combustible, its performance in the event of a fire is poor and it is likely to buckle and fail. The upshot of this is that where a steel-framed structure is used, some form of fire protection will need to be applied or used to encase the steel members. This may include a concrete or masonry surround, sprayed applications of cement and vermiculite, plasterboard casing or specialist paint applications. Steel members are manufactured in standard shapes to suit different loading requirements (universal columns, universal beams, channels, angles, etc.), and each of these standard shapes is manufactured in a range of standard sizes.

Structural steelwork is classified for the purposes of measurement under NRM2 work section 15. Two scenarios are provided in this work section – a framed construction (NRM2 15.1/2) and an isolated structural member (NRM2 15. 3/4). Considering the framed construction first, NRM2 recognises two distinctive cost components associated with framed structural steelwork – fabrication (making the steel) and erection (assembling the frame on site). Fabrication would normally take place at a steel fabrication yard and involves cutting pieces of steel and connecting them together ready for assembly on site. The component parts are then transported from the fabrication yard to site where they are positioned, aligned and secured to form a complete frame on prepared foundations.

In each case (fabrication and erection) the unit of measurement is tonnes (t). Even so, it is worth noting that the initial booked dimensions are recorded in linear metres rather than in the tonnes that will appear in the eventual Bill of Quantities (BQ). The lengths of universal beams, universal columns etc. are then converted from metres to tonnes using standard steel weight tables. These work on the principle that a specific type and size of steel member has a known weight for every metre of its length, which will be expressed in kg/m (Figure 8.1).

As with other work sections, there is a requirement that the measured items are accompanied by the following minimum information: general arrangement drawings, the relative position of all members together with their types and respective size, and details of connections.

8.02 Isolated structural members

Before looking at a suggested approach to measuring a structural steel frame, as a first step it would be sensible to consider an isolated steel member. Typically this would be required where an opening in a load-bearing wall is required. A structural engineer or building surveyor will provide the specification for this, identifying the size, length and the type of isolated steel beam necessary. So far as measurement is concerned, all that is required at this stage is for the length of the isolated steel beam to be booked in the dimension column with the type of steel member, together with the cross-sectional dimensions and weight expressed in kilograms per metre, written alongside this in the description column (see Figure 8.1). As noted earlier, the unit of measurement for steelwork is tonnes (recorded to two decimal places of a tonne in the BQ). In order to achieve this, the measurer will need to convert the recorded length of steel into a weight. Since we already know from the design information available the weight of the steel (expressed in kg/m), all that is required is for this length to be multiplied by the weight per metre of the steel. This can be set up on a sheet of dimension paper as shown in Figure 8.1.

8.03 Framed steelwork

Before measuring framed structural steelwork, it is necessary to have some understanding of the various fittings and components which form part of the finished structure. For the purposes of identification, these are components that enable steel members to be joined together and would typically include brackets, angle cleats and splicing plates (see Figure 8.2).

The need for a systematic approach should be self-evident since it is very easy to measure, say, an angle cleat once with the beam (to which it is attached) and then once more to the column (to which it is also attached). These additional components share the same unit of measurement as any other structural metalwork (tonnes).

In regard to booking dimensions for fittings, NRM2 offers two approaches. The first would require calculating the weight of each individual fitting (see Figure 8.3) and then adding all of these together, thereby arriving at a total calculated weight for all fittings (NRM2 11.5.1). Although this sounds a daunting task, as we will see later this is not as onerous as it at first appears. Alternatively, the weight of all fittings can be based on a percentage of the total

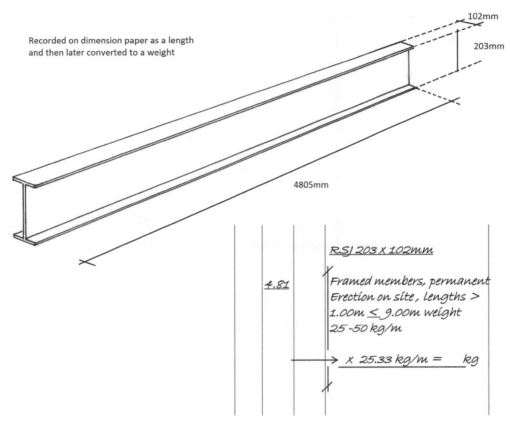

Recorded on dimension paper as a length
and then later converted to a weight

102mm

203mm

4805mm

RSJ 203 x 102mm

4.81

Framed members, permanent
Erection on site, lengths >
1.00m ≤ 9.00m weight
25-50 kg/m

x 25.33 kg/m =   kg

Figure 8.1 Rolled Steel Joist (RSJ).

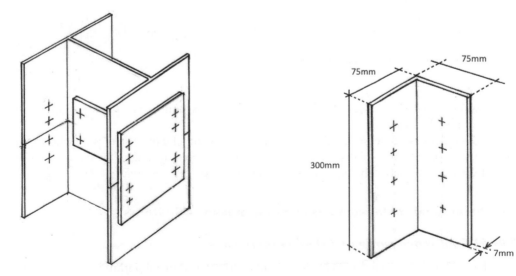

75mm

75mm

75mm

300mm

7mm

Figure 8.2 Examples of framed steel fittings.

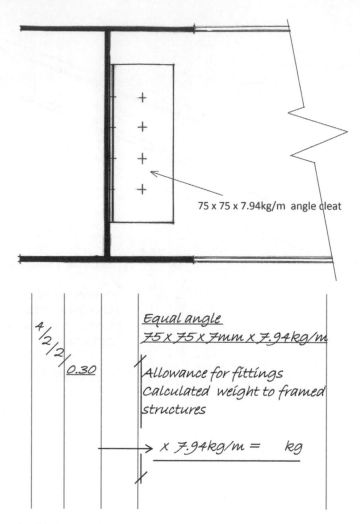

Figure 8.3 Example of fitting (equal angle used as angle cleat).

frame weight (NRM2 11.5.2). Typically this would be around 5% of the weight of all structural metal members.

For the purposes of measuring fittings where the mass or weight of steel is not shown on the drawings or readily available elsewhere, it can be assumed to weigh 785 kg/m² per 100 mm of thickness, which is the equivalent of 7.85 tonnes per m³ (see Figure 8.4).

8.04 Structural metalwork: associated measurable items

8.04.01 Cold-rolled purlins + EO for cranked ends

In order to support wall and roof cladding systems, cold-rolled metal purlins or cladding rails are fixed at ninety degrees to the main structural steel frame of a building. These are

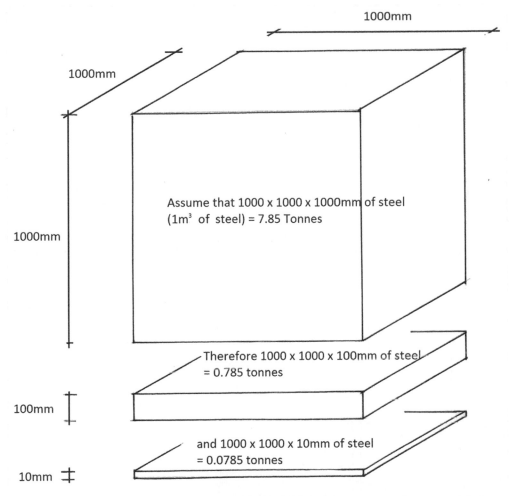

Figure 8.4 Framed steelwork: Establishing the weight of steel fittings/components where no weight is given.

manufactured in 'C-' or 'Z'-sections from cold-formed, pre-galvanised steel channels and are measured in linear metres giving the type, size, method of fixing and any proprietary reference in the description. The description should also identify whether these are purlins/ cladding rails, sag rods, stays or other type of member (NRM2 15.6.1/2.1–4). Where purlins or cladding rails are designed to include 'cranks', these should be enumerated separately, giving the dimensions of the member in the description (NRM2 15.7.1.1).

8.04.02 Profiled metal decking + EO

Profiled metal decking consists of interlocking profiled sheets of galvanised metal that provide a base for an in situ concrete floor and are laid directly onto a steel or concrete frame. It should be noted that these could also be measured and described as permanent formwork

(NRM2 11 Formwork note 4). The distinction that determines whether this is measured with structural metalwork or as part of NRM2 work section 11 is based on how the project is packaged.

8.04.03 Holding-down bolts

Pad foundations would be the usual substructure designed to support each individual column of a structural steel frame. Another way to describe this would be as the junction or interface between the steelwork superstructure and the concrete substructures. A welded or bolted steel plate, normally with four preformed bolt holes, is fixed to the base of each steel column. This in turn is bolted down (held down) through the concrete foundations by four 'holding-down' bolts. The bolts for each column would need to have been cast into the concrete pad foundation as part of the substructure construction. The 'holding-down bolts' sit with their threaded ends projecting vertically from the foundation and align with the preformed bolt holes in the steel base plate. As each steel column is craned into position, it will be fixed with steel washered nuts to the threaded end of the projecting holding-down bolts (see Figure 8.5).

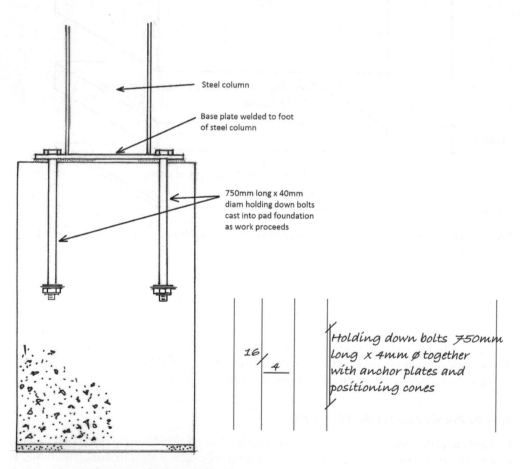

Figure 8.5 Section through base plate and holding-down bolts.

Holding-down bolts are measured as enumerated items, stating the type and diameter in the description. Associated anchor plates and expanded polystyrene cone mouldings are also included in the description (NRM2 15.10.1.1.1). The measurer should be clear as to whether the holding-down bolts are measured as part of the structural metalwork (as described above) or as part of the substructure in situ concrete works (NRM2 11.41.1.1).

Special bolts are measured in the same fashion, as enumerated items giving their type and diameter in the description. Special bolts are defined as any other fixings other than holding-down bolts or grade 4.6 black bolts used to secure the main steel framework.

8.04.04 Surfaces treatments

As the title suggests, this is a protective/decorative finish to the structural steelwork/metalwork that forms the frame of the structure. The description should include the type of finish or coating, including galvanising, sprayed coatings, painting or other treatments, and should be measured in square metres. Booked dimensions are based on the length and girth of each of the structural steel members, and there is an assumption that the same finish is applied

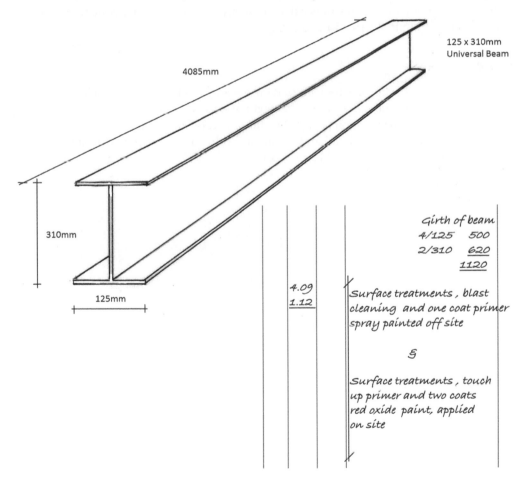

Figure 8.6 Surface treatments and painting.

to the entire frame (see Figure 8.6 for the necessary waste/girth calculation followed by recorded dimensions for surface treatments). Mention in the description also needs to be made of whether the surface treatment takes place on or off site. In many cases it is likely that a protective base application of red oxide primer (or galvanising) will be carried out as part of the manufacturing process (off site), while finishing coat/s are more likely to be applied once the structure has been erected (on site). Any preparatory work, the number and thickness of coats, the fire rating and the type of finish should be stated in the description (NRM2 15.15.1–4.1–2.1–5). This is a departure from the previous version of the Standard Method of Measurement (SMM7), where there were two distinctive items of measurement. The first was 'Surface Preparation' (e.g. sand blasting) and the second 'Surface Treatments' (e.g. painting or galvanising). When using NRM2, 'Surface Preparation' is deemed included.

8.04.05 Isolated protective coatings

Where protective coatings are applied to one-off components or particular zones of the structural steelwork (e.g. to any fire escapes), these are described in similar fashion but should be booked as enumerated items rather than as an area, giving the approximate size or area in the description (NRM2 15.16.1.2–3.1–2).

8.04.06 Testing

Once the structural frame has been erected on site and is substantially complete, it may be the subject of a load test and/or fire protection test or some other type of testing. This will be evident from the specification provided by the structural engineer. Where this is the case, this can be measured as an 'Item' and accompanied by an appropriate description for the type and function of the test (NRM2 17.1–3.1).

8.05 Structural steel frame drawing and specification

This involves SSFrame/05/01 and associated specification notes.

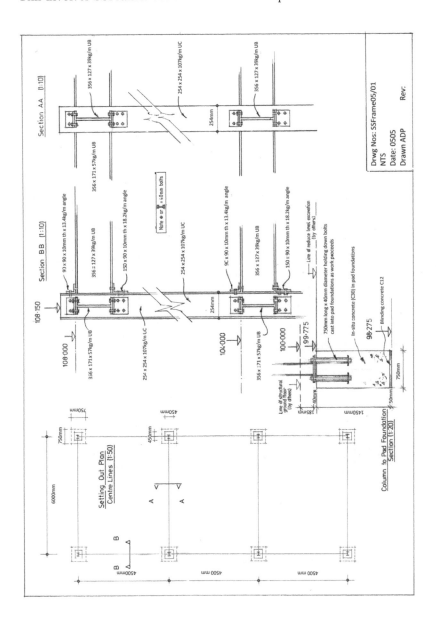

8.06 Measurement example Job ref: SS Frame

Structural steel frame 6A   Job ref: SSFrame   Page 01 of 5

Mandatory information to be provided for structural metalwork :-

See NRM2 work section 15 for full details of mandatory information to be provided .

The work comprises the construction of new build two storey steel frame. All steel members in accordance with BSEN1304 blast cleaned and connected with grade 4.6 blackbolts including test erection and delivery to and offload at site. .

The position , size, type and grade of all members, together with the specification describing fabrication, welding, testing, erection and permissible lifting loads in relation to cranage all in accordance with Structural Engineer's drawing ref SSFrame05/01

Groundwork and concrete work items associated with steel column bases are included in the following sample take-off.

Take-off list:-

SUBSTRUCTURES ⎡ Excavate pits
⎢ Disposal exc mats
⎢ Blinding concrete
⎢ Concrete in pad foundations
⎣ Holding down bolts

STRUCT METALWORK ⎡ Framing fabrication
⎢   columns
⎢   beams (1)
⎢   beams (2)
⎢ Holding down plate assembly
⎢ Framing erection
⎢ Allowance for fittings
⎢ Surface treatments
⎣ Testing

<u>Pad foundations</u>

<u>(Include details of mandatory information for substrata, groundwater etc. as all Chapter 5 headings)</u>

Depth of pad founds

	1 450
<u>Add</u> blinding	<u>50</u>
	<u>1 500</u>

8	0.75		
	0.75	Excavating commencing at reduce level (99.775) foundation excavation not exceeding 2.00m deep	NRM2 5.6.2.1
	1.50		

&

		Disposal excavated materials off site 10km from site	NRM2 5.9.2.1

8			
	0.75	Plain in-situ concrete (C12) in horizontal work ≤ 300 mm in blinding poured on or against earth or unblinded hardcore	NRM2 11.2.1.1.1
	0.75		
	0.05		

8			
	0.75	Plain in-situ concrete (C30) in mass concrete in pad foundations	NRM2 11.1.1.3
	0.75		
	1.45		

8			
	4	Holding down bolts 750mm long x 40mm Ø together with anchor plates and positioning cones	NRM2 15.10.1.1

Note:-
Supply and fix of base plate and holding down bolts by steel fabricator.

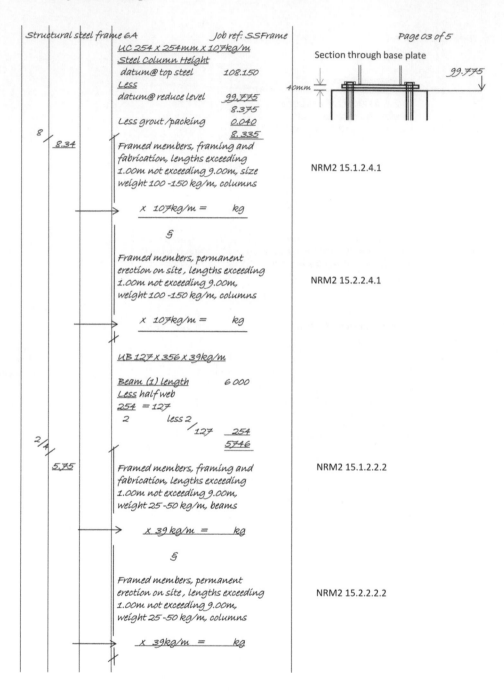

UC 254 x 254mm x 107kg/m
Steel Column Height
datum @ top steel   108.150
Less
datum @ reduce level   99.775
                  8.375
Less grout/packing   0.040
                  8.335

8 / 8.34

Section through base plate
40mm
99.775

Framed members, framing and
fabrication, lengths exceeding
1.00m not exceeding 9.00m, size
weight 100 -150 kg/m, columns   NRM2 15.1.2.4.1

x 107kg/m =   kg

&

Framed members, permanent
erection on site, lengths exceeding
1.00m not exceeding 9.00m,
weight 100 -150 kg/m, columns   NRM2 15.2.2.4.1

x 107kg/m =   kg

UB 127 x 356 x 39kg/m

Beam (1) length   6 000
Less half web
254 = 127
2   less 2
        / 127   254
              5746

2/A / 5.75

Framed members, framing and
fabrication, lengths exceeding
1.00m not exceeding 9.00m,
weight 25 -50 kg/m, beams   NRM2 15.1.2.2.2

x 39 kg/m =   kg

&

Framed members, permanent
erection on site, lengths exceeding
1.00m not exceeding 9.00m,
weight 25 -50 kg/m, columns   NRM2 15.2.2.2.2

x 39kg/m =   kg

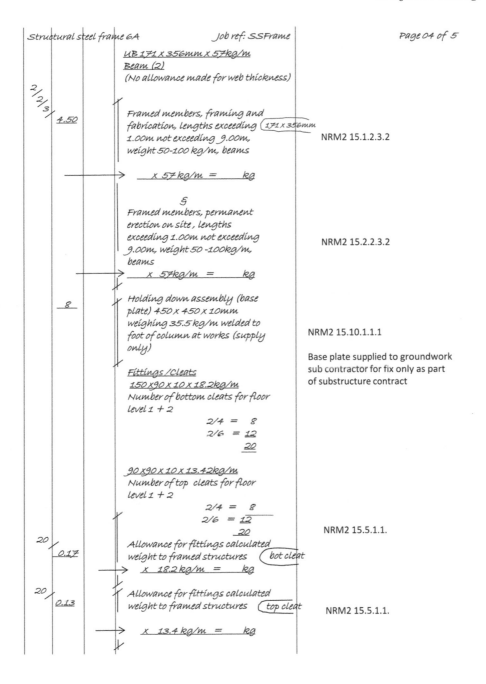

UB 171 x 356mm x 57kg/m
Beam (2)
(No allowance made for web thickness)

2/2/3/ 4.50

Framed members, framing and
fabrication, lengths exceeding (171 x 356mm)
1.00m not exceeding 9.00m,
weight 50-100 kg/m, beams   NRM2 15.1.2.3.2

x 57 kg/m =   kg

&

Framed members, permanent
erection on site, lengths
exceeding 1.00m not exceeding
9.00m, weight 50 -100kg/m,   NRM2 15.2.2.3.2
beams
x 57kg/m =   kg

8   Holding down assembly (base
plate) 450 x 450 x 10mm
weighing 35.5 kg/m welded to
foot of column at works (supply   NRM2 15.10.1.1.1
only)

Base plate supplied to groundwork
sub contractor for fix only as part
of substructure contract

Fittings /Cleats
150 x90 x 10 x 18.2kg/m
Number of bottom cleats for floor
level 1 + 2
                  2/4 =   8
                  2/6 = 12
                         20

90 x90 x 10 x 13.42kg/m
Number of top cleats for floor
level 1 + 2
                  2/4 =   8
                  2/6 = 12
                         20   NRM2 15.5.1.1.

20/0.17   Allowance for fittings calculated
weight to framed structures (bot cleat)
x 18.2 kg/m =   kg

20/0.13   Allowance for fittings calculated
weight to framed structures (top cleat)   NRM2 15.5.1.1.
x 13.4 kg/m =   kg

172   *Steel-framed buildings*

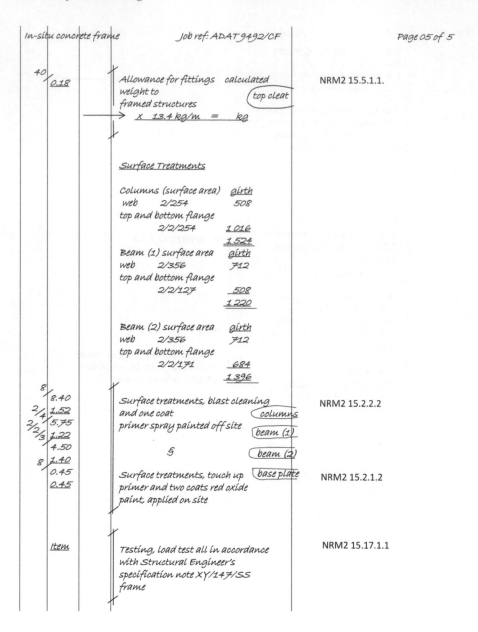

40	0.18	Allowance for fittings calculated weight to framed structures	NRM2 15.5.1.1.

top cleat

x 13.4 kg/m =   kg

Surface Treatments

Columns (surface area)	*girth*
web 2/254	508
top and bottom flange	
2/2/254	1 016
	1 524
Beam (1) surface area	*girth*
web 2/356	712
top and bottom flange	
2/2/127	508
	1 220
Beam (2) surface area	*girth*
web 2/356	712
top and bottom flange	
2/2/171	684
	1 396

8	8.40	Surface treatments, blast cleaning and one coat primer spray painted off site	NRM2 15.2.2.2
2/4	1.52		
2/2/3	5.75		
	1.22		
	4.50		
8	1.40		

columns

beam (1)

&

beam (2)

	0.45	Surface treatments, touch up primer and two coats red oxide paint, applied on site	NRM2 15.2.1.2
	0.45		

base plate

Item	Testing, load test all in accordance with Structural Engineer's specification note XY/147/SS frame	NRM2 15.17.1.1

9 Structural timber

9.01 Introduction

Timber has been used as a structural part of buildings since time immemorial. Compared with other structural materials (concrete and steel) it is not particularly strong but this is compensated by its light weight, which allows for large timbers to be employed without excessive increase in the weight of the structure.

Increasingly nowadays, timber components are prefabricated: for pitched roofs of a domestic scale there are few options – trussed rafters dominate. A comparison of cost between a traditional cut roof and a trussed rafter roof shows little difference in the cost of supply, but the trussed roof is reputed, on average, to be four times as fast to erect.

The process of prefabrication has been taken one stage further with the development of timber-framed housing, which now claims 25% of the UK 'new build' housing market. Given enhanced energy use and performance, structural insulated panels (SIPS) can offer higher levels of thermal insulation than can be achieved with traditional masonry construction.

9.02 Units of measurement and standard sectional timber sizes

The unit of measurement for structural timber naturally falls between enumerated items (trusses) and linear metres (joists, rafters, studs). In the case of the latter, the description must include the nominal cross-sectional dimensions of the timber being measured, while the former must fully describe the engineered or prefabricated member required and this is undoubtedly best communicated with the aid of a dimensioned description.

Timber used in the construction industry is sawn at the mill into standard sectional sizes, which are known as 'nominal' or 'basic' sizes (see Table 9.5). The process of sawing leaves an irregular or rough face on the surface of the timber; this gives rise to the term 'sawn'. Most primary or structural timbers are supplied sawn and NRM2 deems (assumes) that all timber is expressed in a sawn or nominal size (see introductory clauses for NRM2 work section 16 Carpentry; works and materials deemed included). Where a smooth or planed wrought (wrot) finish is required, the dimensioned description should be suffixed with the word 'finished' or 'fin'. It is unlikely that this will apply to many structural timbers (other than trusses), although some first-fix items (e.g. fascia and soffit boards) are grouped for measurement purposes with this structural section of NRM2 (NRM2 16.4). On the few occasions where a wrot finish is required, NRM2 clause 16.4.*.*1 allows for either a linear or superficial measurement depending on the width or girth of the planed finish required.

A more detailed explanation of the sizing and measurement of non-structural timber components can be found in Chapter 11, Windows, doors and joinery.

9.03 Measurement approach

Drawn details identifying the scope and location of the work should be provided in order to satisfy the opening conditions of NRM2.16. It is also important to include details of the kind and quality of timber, together with the method of fixing (where not at the discretion of the contractor) and any surface treatments that are applied as part of the production process. These, along with other similar details, are listed under the heading Mandatory Information to be provided at the commencement of NRM2.16. To save unnecessary repetition these may be included in a general heading to the take-off, as shown in Figure 9.1.

The classification table given in NRM2.16 discriminates between engineered or pre-fabricated members/items (NRM2.16.2) and primary or structural timbers (NRM2.16.1). The former includes what would traditionally be described as 'loose timbers' such as rafters, floor joists, wall plates, stud partitions and wall members, while the latter covers roof trusses, portal

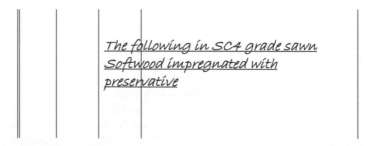

Figure 9.1 Specification details of the timber can be included as part of a description heading to avoid unnecessary repetition.

frames, open web joists and wall panels. Many projects will include a combination of both. It is important to determine which is which.

The following techniques are commonly used in order to establish the number and length of timber components, and a brief résumé of these techniques and approaches follows.

9.03.01 Calculating the number of joists

Drawn information provided by architects rarely includes a floor or roof plan that identifies the number of joists, rafters or trusses required. Instead, the measurer must establish these from the other information available on the drawing. This will include the spacing centres of joists or rafters (commonly 400, 450 or 600 mm), together with external or internal plan dimensions and an indication on the drawing of the span direction (which would normally be the shorter of the room plan dimensions), and is usually represented by a directional arrow.

The stages involved in establishing the number of joists are now listed (Figures 9.2 and 9.3):

1   Find internal length of room.
2   Find centre-to-centre dimension of first and last joists.
3   Divide dimension found in (2) above by the joist centres to establish the number of *spaces*.
4   Round answer up to whole number (on rare occasions the answer will already be a whole number).
5   Having found the number of *spaces*, add one to obtain the number of *joists*.

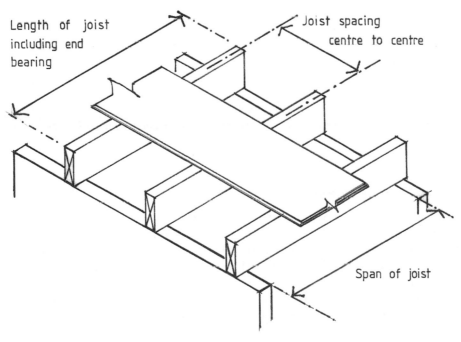

Figure 9.2 Isometric of floor joists shown as a visual aid to calculating the number of joists (or rafters) required.

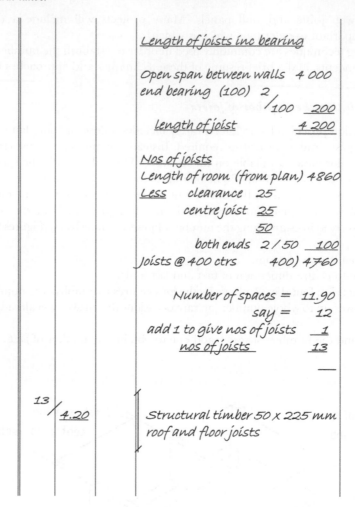

Figure 9.3 Sample waste calculations and booked dimensions for establishing number of joists (or rafters) required.

9.03.02 Calculating the number of rafters

A similar approach should be adopted to establish the number of rafters in a hipped roof.

The inclusion of hips and valleys can be ignored when measuring common rafters, and no distinction is made between hipped and gable ended roofs. The total length of jack-rafters at a hip or valley is nominally equal to the length of the corresponding common rafters in a gable roof (see Figure 9.4).

The combined length on the plan of each pair of jack-rafters is equivalent to one common rafter. However, it will be necessary to take one extra rafter at each end on the same line as the ridge, as detailed in Figure 9.4.

The waste calculation used to identify the number of joists can also be used to establish the number of trussed rafters, cut rafters in a gable roof, or studs in partition walls.

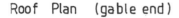

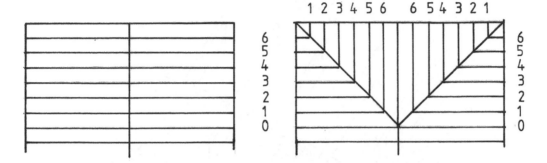

Figure 9.4 Plan showing a comparison between gable end and hipped end roof timbers illustrating an approach to identifying the number of rafters required.

9.03.03 Establishing the length of structural timbers

A further waste calculation will be necessary to establish the length of structural timbers. In the case of floor or ceiling joists this is reasonably straightforward, and simply requires the addition of end bearing to the clear span dimension (see Figures 9.2 and 9.3).

Alternatively, where the joist is suspended in joist hangers, the measured length of timber would simply be the clear span. In this case an inclusion must also be made for joist hangers, and these will be enumerated in accordance with NRM2 16.6.2.7.

A similar approach may be adopted for the vertical studs in a stud partition wall, with appropriate adjustment for head and sill timbers. The length of rafters can be established in a number of ways, as shown in Figures 9.5, 9.6 and Table 9.1.

Table 9.1 Secants of common roof slopes.

Roof slope	Secant
17½°	1.049
22½°	1.082
30°	1.155
35°	1.221
37½°	1.26
40°	1.305
45°	1.414

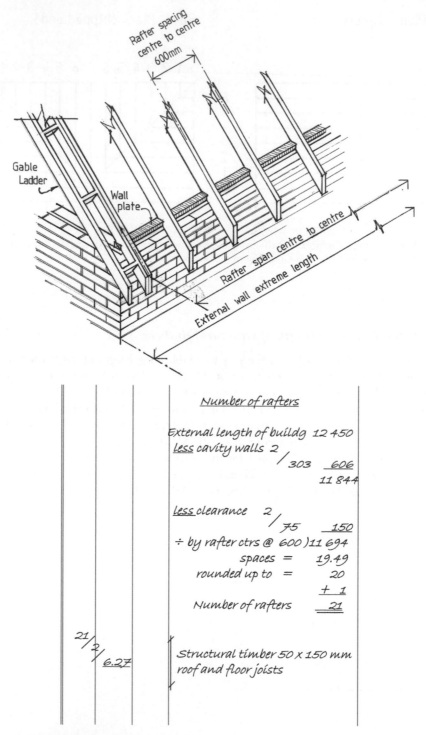

Figure 9.5 Establishing the length of rafters.

Length of rafter can be found by
using one of the following methods :-

1. Scale length from section

2. Use Pythagoras theorem $L = \sqrt{a^2 + b^2}$

3. Use Natural Secant $L = a \times \sec 45°$

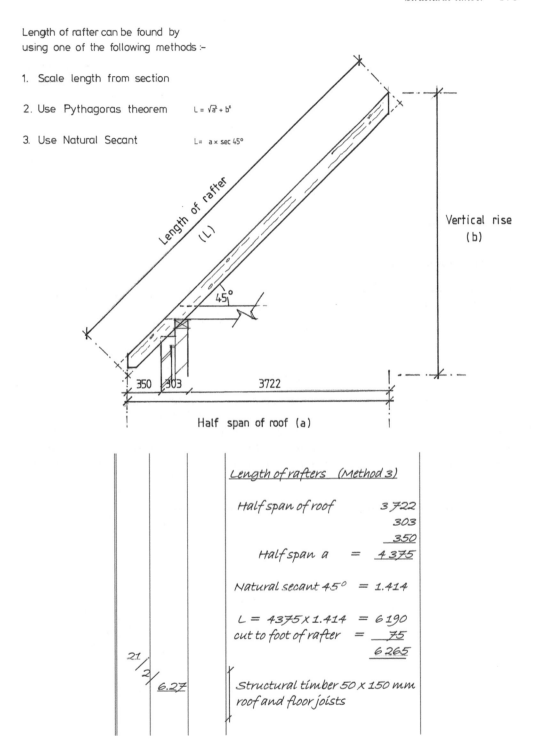

Length of rafter (L)

45°

Vertical rise (b)

350   303   3722

Half span of roof (a)

<u>Length of rafters   (Method 3)</u>

Half span of roof   3 722
                                       303
                                       350
    Half span a   =   4 375

Natural secant 45° = 1.414

L = 4375 × 1.414 = 6 190
cut to foot of rafter =   75
                                      6 265

Structural timber 50 × 150 mm
roof and floor joists

21
/2
/ 6.27

Figure 9.6 Establishing the length of rafters.

9.03.04 Length of hip and valley rafters

Figure 9.7 shows a plan view of a hipped roof. The length of a hip or valley cannot be scaled directly from plan since the length required is the length on slope. NB: In order to establish this using the following technique you will need a protractor, a scale rule and a pencil. The following approach is proposed:

1   Set out the roof height as shown in Figure 9.7, at right angles to the hip.
2   Scale off the vertical height of the roof along this line.
3   Join the two lines to form a right-angled triangle and scale the length of the hip from the drawing.

Establishing the length of rafters may be determined in a number of ways (method 3 is illustrated in Figure 9.6).

9.03.05 Complex roof shapes

When faced with a complex roof plan, it is suggested that the various component parts of the roof are subdivided into their basic component units. The intention would be to identify the principal roof as the main component followed by the addition of any projections (see Figure 9.8).

9.04 Presentation in the Bill of Quantities

The descriptive part of the measurement was partly covered in the introduction to this chapter. For traditional, rather than prefabricated, components, NRM2 groups together a variety of structural timber members based upon their location and function (NRM2 16.1.1.1–8). The length of the timber being measured is recorded in the dimension column, and the timesing column is used to multiply this by the number of times the particular component occurs. The

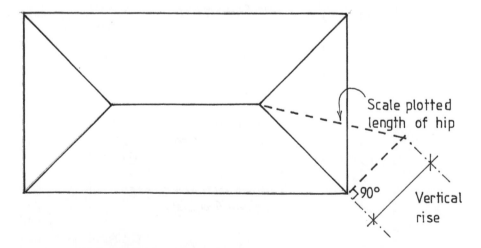

Figure 9.7 Calculating the length of hip and valley rafters.

More complicated roof layouts

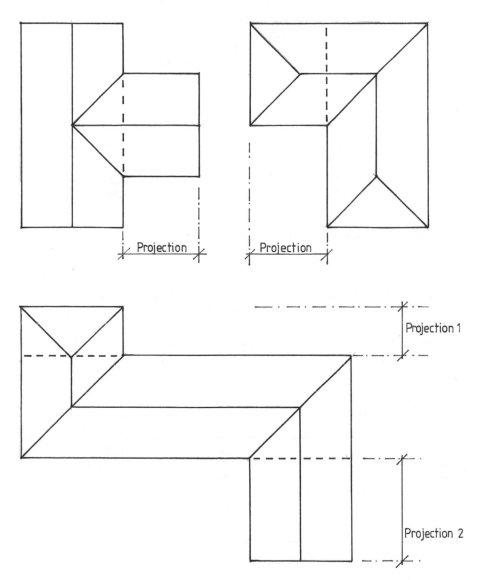

Figure 9.8 Establishing the area of complex roof plan shape.

BQ will of course only record a single total length, accompanied by the description that identifies the type of member and the cross-sectional dimension.

A completed description for ceiling joists might be presented as shown in Figure 9.9. A similar layout and presentation should be adopted for any other structural timber items. Note the use of signposting to identify, in this instance, ceiling joists. The preparation of the BQ will amalgamate all floor and roof joists of the same sectional size together regardless of

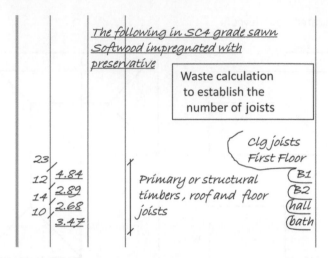

The following in SC4 grade sawn
Softwood impregnated with
preservative

Waste calculation
to establish the
number of joists

Clg joists
First Floor

23
12   4.84   B1
14   2.89   Primary or structural   B2
10   2.68   timbers, roof and floor   hall
      3.47   joists   bath

Figure 9.9 Completed dimensions and descriptions for joists.

whether they are associated with a floor or a pitched roof (NRM2 16.1.1.4). There will be no obvious way to distinguish a roof joist from a ceiling joist once these items have been abstracted and billed, apart from signposts included by the measurer at the time of the take-off.

The take-off lists given in Tables 9.2, 9.3 and 9.4 are offered for general guidance only. Since every situation will vary, it is unlikely that the construction will exactly match this.

Table 9.2 Take-off list 1.

Item	Unit	*NRM2 ref*
Wall plates	m	16.1.1.3
Floor joists	m	16.1.1.4
Timber beams	m	16.1.1.5
Joist hangers, metal straps and connectors	nr	16.6.1 or 2.1-8
Joist strutting	m	16.1.1.8
Insulation	m^2	31.3.1.1-3.1
Boarding (flooring)	m/m^2	16.1/2.1.1-3

Table 9.3 Take-off list 2.

Item	Unit	*NRM2 ref*
Wall plates	m	16.1.1.3
Floor joists	m	16.1.1.4
Timber beams	m	16.1.1.5
Joist hangers, metal straps and connectors	nr	16.6.1 or 2.1-8
Joist strutting	m	16.1.1.8
Insulation	m^2	31.3.1.1-3.1
Boarding (flooring)	m/m^2	16.1/2.1.1-3

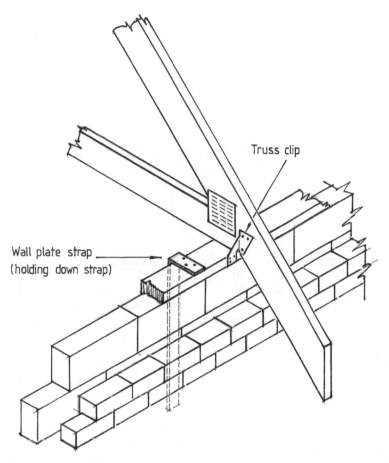

Figure 9.11a Isometric detail of the junction between roof structure and external cavity wall showing metalwork used to secure truss to wall plate and wall plate to internal face of cavity wall.

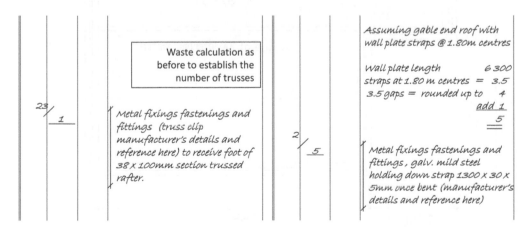

Figure 9.11b Booked dimensions for metalwork associated with trussed rafter and wall plate (see Figure 9.11a).

9.06 Drawings and example take-off

9.06.01 Intermediate timber floor drawings and specification

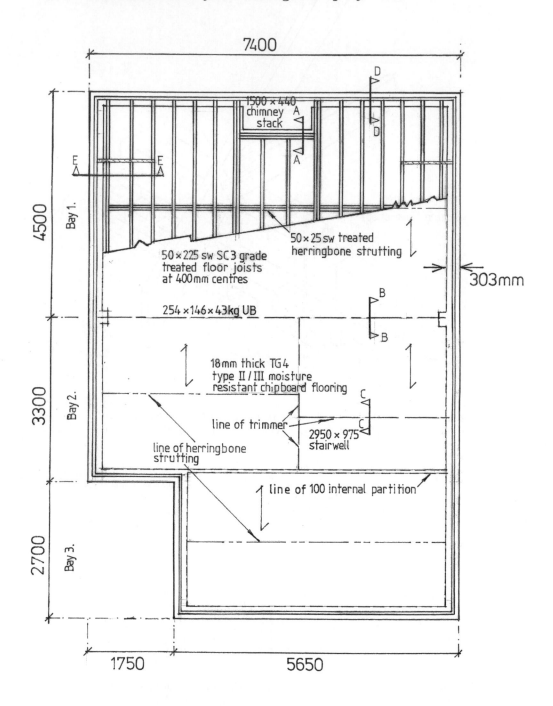

Suspended Timber Intermediate Floor
Section & Details

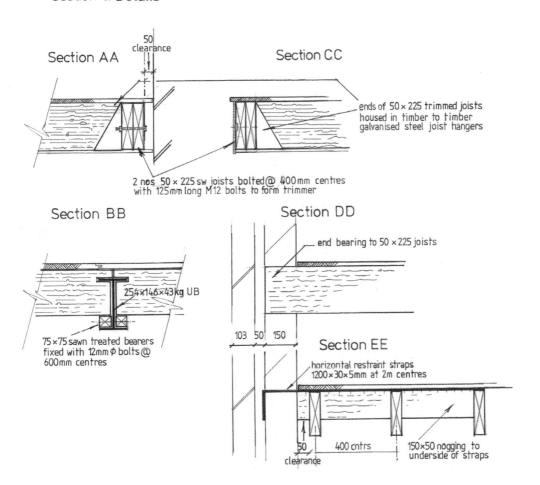

Section AA

50 clearance

Section CC

ends of 50 × 225 trimmed joists housed in timber to timber galvanised steel joist hangers

2 nos 50 × 225 sw joists bolted @ 400mm centres with 125mm long M12 bolts to form trimmer

Section BB

254×146×43kg UB

75×75 sawn treated bearers fixed with 12mm ⌀ bolts @ 600mm centres

Section DD

end bearing to 50 × 225 joists

103 50 150

Section EE

horizontal restraint straps 1200×30×5mm at 2m centres

50 clearance

400 cntrs

150×50 nogging to underside of straps

9.06.02 Intermediate timber floor measurement example

Job ref: BMMS 7476/HH.

Suspended Timber Intermediate floor   Job ref: BMMS 7476/HH   Page 01 of 10

The work comprises the construction of a suspended timber intermediate floor bearing on an isolated structural steel Universal Beam for a residential two storey property all in accordance with the floor plans, sections and details provided.

Take off list

SUSPENDED TIMBER STRUCTURAL FLOOR

METALWORK
- Isolated structural member (Universal Beam)
- Surface Treatments to UB
- Padstones
- Timber bearers to UB
- Bolts

STRUCTURAL TIMBERS
- Joists
- Strutting to joists (herringbone)
- Galvanised metal restraint straps
- Strutting to straps

ADJUSTMENT FOR CHIMNEY STACK
- Trimming to stack
- Joist hangers
- Bolts

ADJUSTMENT FOR STAIRWELL
- Trimming to stack
- Joist hangers
- Bolts

FLOOR BOARDING
- Chipboard floor boarding
- Sound deadening insulation

Suspended Timber Intermediate floor   Job ref: BMMS 7476/HH   Page 02 of 10

<u>Universal Beam</u>

Length	7 400
<u>Less</u>	
Cav walling 2/303	<u>606</u>
projection both sides	6 794
<u>Add</u>	
end bearing 2/150 =	<u>600</u>
	<u>7 094</u>

<u>Metalwork</u>

7.09   Isolated structural member,   NRM2 15.3.1.1
        plain member universal beam
        254 x 146 x 43kg

        x   43kg/m =   kg

<u>Decoration to Universal Beam</u>
Girth 2/2/ 146   584
      2/2/ 254   <u>508</u>
                        <u>1 092</u>

7.09   Prepare with wire brush, two   NRM2 29.3.2.2
1.09   Coats zinc phosphate primer
        to structural metalwork >
        300 mm girth, external,
        application on site prior to fixing

<u>Padstones</u>

2   Precast concrete goods (1:3:6 -20   NRM2 13.1.1.3
        mm agg) padstones 328 x 328
        x 150 mm deep bedded in gauged
        mortar (1:1:6)

190   *Structural timber*

		Suspended Timber Intermediate floor	Job ref: BMMS 7476/HH	Page 03 of 10

2 / 6.79	75 x 75 treated sw first fix battens, bolt fixed through U Beam (bolts measured sep)	NRM2 16.3.1.1

Nos of bolts required

Length of batten 6 794
Bolts @ 400 mm ctrs

$$\frac{6\,794}{400} = 16.98$$

say   17
+   1
18

18	Metal fixings, fastenings and fittings 12mm Ø 200 mm Long mild steel coach bolts	NRM2 16.6.1.6

Length of joists

Bay 1   4 500
less 103
   50   153
     4 347

Bay 2   3 300
less 103
   50   153
     3 147

Bay 3   2 700
add bearing   100
     2 800
less 103
   50   153
     3 147

Suspended Timber Intermediate floor   Job ref: BMMS 7476/HH   *Page 04 of 10*

<u>Nos of joists Bay 1 and 2</u>

		7 400
<u>less</u> walls	2/303	606
		6 794
<u>less</u> clearance	2/50	100
		6 694
Less half width of joist	2 / $\frac{1}{2}$ / 50	50
@ 400 mm ctrs	) 6 644	
=		16.61
Rounded to		17
+ 1		
=		18

<u>Nos of joists Bay 3</u>

		5 650
<u>less</u> walls	2/303	606
		5 044
<u>less</u> clearance	2/50	100
		4 944
Less half width of joist	2 / $\frac{1}{2}$ / 50	50
@ 400 mm ctrs	) 4 894	
=		12.23
Rounded to		13
+ 1		
=		14

	Suspended Timber Intermediate floor	Job ref: BMMS 7476/HH	Page 05 of 10

The following in SC3 grade
Sawn softwood impregnated
with preservative

18/	4.35	Structural	(bay 1
18/	3.15	timbers	(bay 2
14/	2.65	50 x 225 mm	(bay 3
		roof and floor	
		joists	

NRM2 16.1.1.4

Herringbone strutting

Bay 1 span	6 794

Bay 2 span	6 794
Less stairwell opg	2 950
	3 844

Bay 3 span	5 650
less 2/303	606
	5 044

6.79	Structural timbers
3.84	50 x 225 mm
5.04	strutting

NRM2 16.1.1.8

Horizontal restraining straps
O' all length reqd for straps

	4 500
	3 300
2/	2 700
	10 500
=	21 000

Less first to last
both sides
2/2/2 000

	8 000
@ 2 000 ctrs)13 000
=	6 500
Rounded to	7
	+ 1
=	8

Suspended Timber Intermediate floor		Job ref: BMMS 7476/HH	Page 06 of 10

Horizontal restraining straps
(Contd)

8	Metal fixings, fastenings and fittings, galvanised mild steel Vertical restraint strap 1200 x 30 x 50mm

NRM2 16.6.1/2.4

Strutting for restraint straps

Length 2/400	800
clearance	50
Half joist width	25
	875

The following in SC3 grade Sawn softwood impregnated with preservative

8 / 0.88	Structural timbers 50 x 225 mm strutting

NRM2 16.1.1.8

Adjustment for chimney stack opening (1500 x 440)

Trimmed joist lengths

Bay 1 (as before)		4 347
less	440	
clearance	50	
width of trim	100	590
		3 757

Trimming joists length

		1 500
add clearance 2 / 50		100
		1 600

Lateral trimmed joist
length 1600

Plan view of intermediate floor

194 *Structural timber*

<u>Adjustment for chimney stack
opening (1500 x 440) (Contd)</u>

<u>Nos of joists to trim @ stack</u>

Span @ trimming joist 1 600
    ctrs @ 400) 1 600
              =   4
<u>less</u> first and last joist = <u>2</u>
      leaves <u>2 joists</u> remaining

Lateral trimmed joist
length 1600

Longitudinal trimmed joist
length 3757

8/	3.15	<u>Deduct</u> Structural timbers 50 x 225 mm roof and floor joists

NRM2 16.1.1.4

8/	2.07	<u>Add</u> Structural timbers 50 x 225 mm ditto
2/	3.15	
	3.10	

NRM2 16.1.1.4

	2	Metal fixings, fastenings and fittings, galvanised mild steel joist hangers to suit 50 x 225 mm joist

NRM2 16.6.1/2.7

	2	Ditto to suit 100 x 225 mm joist

<u>Establishing nos. of Bolts</u>

@ 400 ctrs <u>on 1600 long</u>
          trimming joist
= 4 joists, therefore = 3 gaps
allow 1 bolt/gap = <u>3 bolts</u>

@ 400 ctrs on <u>4347 long</u> trimmers

<u>Less</u>   4 347
first and last 2/400   <u>800</u>
  bolts @   400) 3 547
      = 8.87 say <u>9 bolts</u>

Suspended Timber Intermediate floor   *Job ref: BMMS 7476/HH*   *Page 08 of 10*

2	3	
	9	

Metal fixings, fastenings and
fittings, mild steel coach bolts
M12 Ø x 125mm long

NRM2 16.6.1/2.6

<u>Adjustment for stairwell opening</u>
<u>(size 2950 x 975)</u>

<u>Trimmed joist lengths</u>
Bay 2 (as before)   3 147
   <u>less</u>   975
      trimmer 2/50   100   <u>1 075</u>
                                <u>2 072</u>

<u>Trimming joists length</u>
                          2 950
   <u>add</u> end bearing   <u>150</u>
                          <u>3 100</u>

Nos of joists to trim @ stairwell
              @ 400 ctrs)<u>2 950</u>
               = 7.375

excluding first and last
   therefore rounded to <u>8 joists</u>

<u>The following in SC3 grade</u>
<u>Sawn softwood impregnated</u>
<u>with preservative</u>

8	
	3.15

<u>Deduct</u>
Structural timbers
50 x 225 mm
roof and floor joists

NRM2 16.1.1.4

8	2.07
2	3.15
	3.10

<u>Add</u>
Structural timbers
50 x 225 mm ditto

NRM2 16.1.1.4

7		Metal fixings, fastenings and fittings, mild steel coach bolts M12 Ø x 125mm long	NRM2 16.6.1/2.7
1		Ditto to suit 100 x 225mm joist	NRM2 16.6.1/2.7

<u>Establishing nos of bolts to trimming joist</u>

Trimming joist	3 147
ditto	3 100
Bolts @ 400 ctrs)6 247	
= 15.6 (say 16)	
rounded to <u>17</u>	

17		Metal fixings, fastenings and fittings, mild steel coach bolts M12 Ø x 125mm long	NRM2 16.6.1/2.6

<u>Floor Boarding</u>

length	4 500
	3 300
	<u>2 700</u>
	10 500
Less 2/ 303	<u>606</u>
	<u>9 894</u>
width	7 400
Less 2/ 303	<u>606</u>
	<u>6 794</u>

Suspended Timber Intermediate floor   Job ref: BMMS 7476/HH   Page 10 of 10

9.89		Boarding, flooring etc. over
6.79		600 mm wide, 18mm thick
		TG4 type II/III moisture
		resistant chipboard flooring
1.75	Ddt	(2440 x 600 mm) to floors
2.70		joints glued and cramped,
2.95		fixed with 40mm long floor
0.98		brads to sw joists, horizontal
1.50		
0.44		&

want

stairwell

chimney
stack

NRM2 16.4.2.1

100 mm thick mineral wool
sound deadening quilt
between joists

NRM2 31.3.1.3.1

9.06.03 Pitched roof structure drawings and specification with measured example (see 10.06 and 10.07)

Job ref: BMMS 7476/HH.

The work comprises the construction of a suspended timber intermediate floor to a residential property, all in accordance with floor plans, sections and details provided

TRADITIONAL PITCHED CUT ROOF

NRM2 requires the following drawings to be available:-
Roof plan/s, principle roof sections, external elevations.

Mandatory information to be provided for carpentry and roof coverings :-
See NRM2 work sections 16 and 18 for mandatory information to be provided .

Take-off list

STRUCTURAL TIMBERS
* Wall plate
* Metal fixings/fastenings
* Ceiling joists
* Rafters
* Ridge Board
* Hip and Valley Board
* Struts
* Binder
* Insulation

Pitched roof structure			Job ref: BMMS 7476/HH	Page 02 of 07

<u>Roof Structure (Contd)</u>

<u>The following in SC3 grade
Sawn softwood impregnated
with preservative</u>

Wall plate Length

2/ 8900		17 800
2/2750		5 500
Ext wall dims		23 300

103 bwk	<u>less</u>		
50 cavity	4/2/153		
153		1 224	
Ext girth of wall plate		22 076	

<u>add</u>	4/2/	
corner o'lap	2/100	800
halved joints	/150	300
		23 176

All labours deemed included but no obvious mention of additional material required in forming half joints

23.18	.Structural timbers 100 x 75 mm sawn sw wall plate bedded in gauged mortar (1:2:9)

NRM2 16.1.1.3.9

<u>Wall plate straps</u>

wall plate girth	22 076

Straps @ 1800 ctrs
22 076 ÷ 1800
  = 12.26 gaps
    say 13 + 1 = <u>14 straps</u>

14	Metal fixings, fastenings and fittings 900 x 30 x 5mm thick holding down strap, screw fixed to timber wall plate and blockwk

NRM2 16.6.1.4.1/2

Roof Structure (Contd)

Nos of ceiling joists; main roof

Main roof run	8 900
less	
Ext wall	288
Clearance	50
half joist width	25
both ends 2 / 363	726
@ 400 ctrs	) 8 174
	= 20.14 gaps

Rounded up to 21 + 1 = 22 joists

Nos of ceiling joists; roof projection

roof projection length	2 750
add eaves proj	250
	3 000
less ext skin wall	103
o'all clear span	2 897
	400) 2 897
@ 400 centres =	7.24 gaps

Rounded up to 8 + 1 = 9 joists

Length of joists
Main roof and projection

=	4 000
less cavity wall 2 /	
288	576
	3 424
add bearing 2 /	
100	200
	3 624

Pitched roof structure			Job ref: BMMS 7476/HHH	Page 04 of 07

<u>Roof Structure (Contd)</u>

<u>The following in SC3 grade
Sawn softwood impregnated
with preservative</u>

22		⌐joists⌐	
9'	3.62	Structural timbers 50 x 100mm rafters and associated roof timbers	NRM2 16.1.1.1.1

<u>Nos of rafters; main roof and
roof projection (See waste
calculation for nos of joists)</u>

<u>Length of rafters</u>
(main roof and projection)

half span of roof	1712
Cavity	288
Eaves overhang	250
half span	2 250
natural secant 45°	x 1.350
	2 936
allow for cut at ridge	100
	3 036

Space for dimension diagram
8.3

2/ 22		⌐rafters⌐	
9'	3.04	Structural timbers 50 x 100mm rafters and associated roof timbers	NRM2 16.1.1.1.1
	3.04	⌐extra rafter at hip end⌐	

<u>Length of ridge</u>
(main roof and projection)

Main roof	8 900
<u>add proj @ verge</u> 2/	
50	100
	9 000
<u>add scarf joint</u> 2/	
175	350
	9 350

Pitched roof structure   Job ref: BMMS 7476/HH

<u>Roof Structure (Contd)</u>
<u>Length of ridge</u>
<u>(main roof and projection Contd)</u>

<u>Projecting hip roof</u>
Half span hip roof

$$\frac{4\,000}{2} \quad = \quad 2\,000$$

<u>add</u>

eaves 2/250	500
roof proj	<u>2 750</u>
	5 250

<u>Less</u>
4 000
eaves 2/250   <u>500</u>

4 500 ÷ 2 <u>2 250</u>

proj roof ridge length   <u>3 000</u>

9.35	Structural timbers
3.00	25 x 175 mm rafters

Structural timbers
25 x 175 mm rafters
and associated
roof timbers   (Main roof) (Proj roof)

NRM2 16.1.1.1

&

Metal fixings, fastenings
and fittings, Uni-Dry stainless
steel ventilating ridge fixing.

NRM2 16.6.1.1

<u>Hip and valley rafter length</u>
Half span (as before)   2 250
Length of rafter   3 036
Slope calc using Pythagoras

$$\sqrt{2.250^2 + 3.036^2}$$

$$\sqrt{14.280} \quad = \quad \underline{3\,780}$$

2/2/ 3.78   Structural timbers
38 x 200 mm rafters   (Hip & valley timbers)
and associated
roof timbers

NRM2 16.1.1.1

Pitched roof structure   Job ref: BMMS 7476/HH

Roof Structure (Contd)

Purlin lengths
main roof   8 900
Less ext wall   103
          cavity   2/ 50
                    153   306
                              8 594
Proj Hip roof purlin   2 750

2/ 8.59	Structural timber (Main roof / Project roof)
2/ 2.75	75 x 225 rafters and associated timbers

NRM2 16.1.1.1

Strut length
Assume required @ mid point
of half span
therefore 1 712 ÷ 2 =   856
Tangent 40° x 856
        0.839 x 856 = 718

4/ 0.72	Structural timbers (struts) 75 x 100 mm rafters and associated timbers

NRM2 16.1.1.1

Binder length   (main roof)

Ceiling joist span
   (centre to centre)   8 174
add   half joist width
      each end 2/ 25   50
   scarf joints 2/100   200
                        8 424

Binder length (proj hipped roof)

Ceiling joist span
   (centre to centre)   2 750
add   half joist width
      each end 2/ 25   50
                        2 800

<u>Roof Structure (Contd)</u>

2/	8.42	Structural timbers	(binder)
2/	2.80	38 x 100 mm rafters and associated timbers	

NRM2 16.1.1.1

<u>Insulation</u>   (main roof)

	length	width
	8 900	4 000
less cav wall		
2 /288	576	576
	8 324	3 324

<u>Insulation</u>   (projecting roof)

<u>length</u>	2750
add eaves	250
cavity wall	288
	3 000

<u>width</u> (as before)   3 324

8.32	Mineral fibre quilt insulation
3.32	100 mm thick laid between
3.00	ceiling joists @ 400mm
3.32	centres, horizontal.

NRM2 31.3.1.3.1

&

NRM2 31.3.1.1.1

Mineral fibre quilt insulation
200 mm thick plain areas,
horizontal.

10 Roof coverings

10.01 Introduction

The measurement of pitched and flat roofs conveniently divides between the roof structure and the roof covering. Many of the waste calculations detailed in the previous chapter (Structural timber) are equally applicable here. In practice, the measurement of roof coverings will follow the measurement of the roof structure and is likely to be carried out by the same person. Eaves and barge boarding, together with guttering and downpipes, are normally included as part of the measurement of roof coverings.

In previous editions of the Standard Method, the trade of Roof Coverings (SMM7) included tiling, slating and the like being grouped with sheet cladding components such as patent glazing and profiled sheet cladding for walls and roofs under Section H Cladding/ Coverings. Asphalt- and felt-covered roofs were included in Work Section J, Waterproofing. NRM2 has adopted the approach of separate work sections for sheet roof coverings (NRM2 17) and tile and slate roof and wall coverings (NRM2 18). The former includes bituminous felt, plastic sheet, sheet metals, rigid board with pre-applied sheet coverings and the like, while the latter covers plain tiling, interlocking concrete tiles, natural and fibre cement slates, natural or artificial stone slating and timber or felt shingles. Mastic asphalt and other liquid applied roofing, tanking and flexible sheet damp proofing systems are included in NRM2 19 Waterproofing.

In all cases, the principal unit of measurement for roof coverings is square metres. All three work sections (NRM2 17, 18 and 19) have a long list of works and materials that are deemed included. The most obvious of these are underlay for sheet roof coverings, and battens and underlay for pitch roof coverings. However, NRM2 18.1/2.1/2.1 appears to contradict the 'deemed included status' of battens and underlay, as the level two heading implies that both underlay and battens should be described. In the preparation of this text, the latter situation has been assumed. This makes for an extended but nonetheless necessary description, which is best set up on dimension paper as a heading (Figure 10.5b). The descriptive part of the measurement must identify the kind, quality and size of materials, together with the method of fixing. The measurement of the main roof slope area should be followed by adjustments for chimneys and dormers. Alternatively this may be included as part of the measurement of

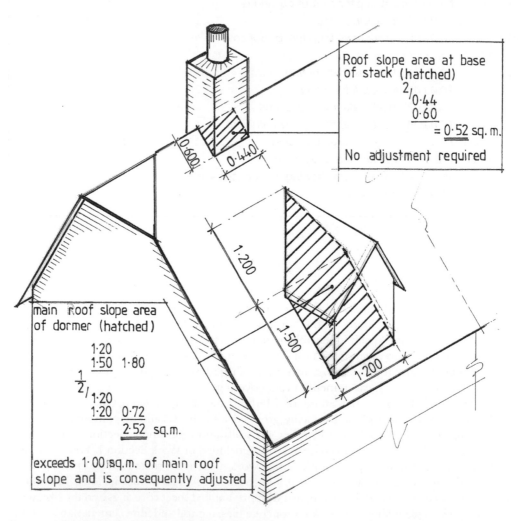

Figure 10.1 Roof area displaced and need for adjustment.

the chimney or dormer. No adjustment is made to the roof covering area for voids of less than one square metre in area (NRM2 17.4.*.*.*.1, NRM2 18.1.*.*.*.1. and NRM2 19. 1.*.*.*.2.). This is illustrated in Figure 10.1, which shows the example of a pitched roof with a chimney stack and a pitched roof dormer. In order to check whether an adjustment in accordance with this last rule is necessary, the affected slope areas are established (see waste calculations as part of Figure 10.1). In the case of the chimney stack, the roof area is 0.52 m^2 so no adjustment is needed. For the dormer, the area displaced is 2.52 m^2 and an adjustment is required.

Ridges, hips, valleys, abutments, eaves, verges and vertical angles are all measured in linear metres and are deemed to include the labour items of cutting, bedding, pointing, forming undercloaks, angles, intersections and preparing ends. Boundary work to voids is measured only where the void area exceeds one square metre.

10.02 Measurement of pitched roof coverings

10.02.01 Roof plan shape

The area of roof coverings is unaffected by the inclusion of hipped ends and valleys, providing the roof pitch remains constant. To illustrate this, Figure 10.2 shows three alternative roof plans all based on the same dimensions and roof pitch. When measured, all three will provide the same roof slope area regardless of whether they are hipped, gabled or a combination of the two.

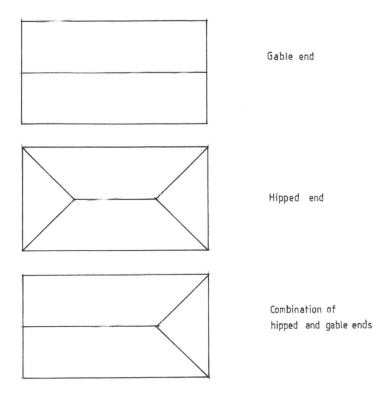

Figure 10.2 Roof slope area unaffected by inclusion of hipped ends and valleys.

In each case the sloping roof area can be established by (initially) ignoring any projections, valleys or hips and simply measuring the main length (L) multiplied by the roof slope (S). Some surveyors choose to enter the dimensions for the roof slope area by recording the plan area of the roof in the dimension column and timesing this by the natural secant of the roof slope in the timesing column. Others prefer to establish the roof slope length as a waste

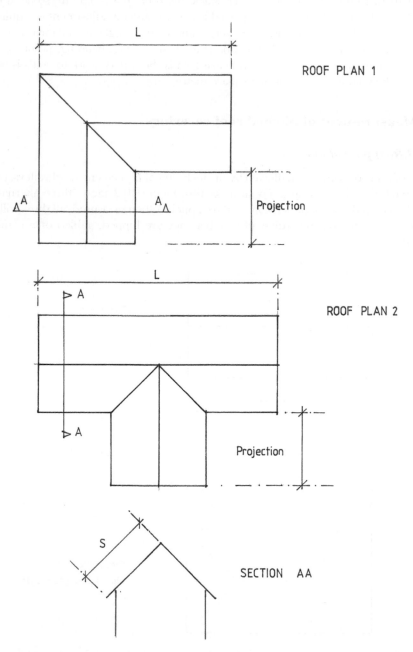

Figure 10.3 Breaking down more complex roof slopes into manageable areas.

calculation first and then record the plan length and roof slope length in the dimension column. This is then timesed by two for each roof slope in the timesing column.

For more complicated roof shapes it is advisable to break these down into a main roof length (L) with the projections (P) (see Figure 10.3).

Where the roof has more than one projection, the procedure should be repeated for each separate projection (Figure 10.4).

If the roof slope of the projection differs from the main roof slope, the initial measurement should be carried out as before, ignoring any projection. After recording the slope area of the projection, an appropriate adjustment should be made for the previous over-measurement (see Figure 10.5a and 10.5b).

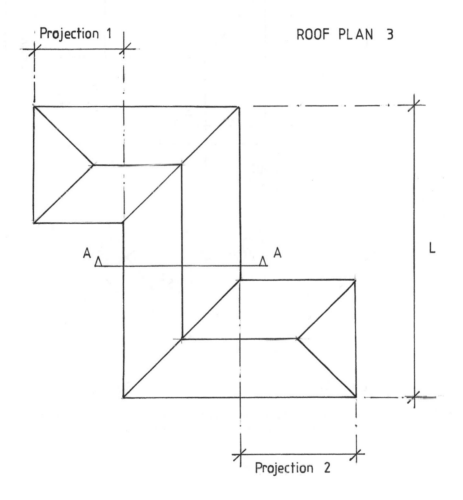

SECTION AA AS ROOF PLANS 1 & 2

Figure 10.4 An approach to measuring multiple projections.

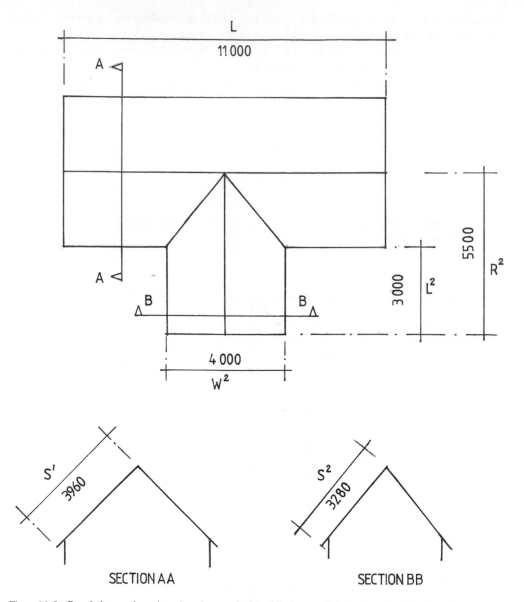

Figure 10.5a Roof plan and section showing roof with differing roof pitch. See the following figure
(Figure 10.5b) for an example of recording dimensions where a single roof has two differing
pitches.

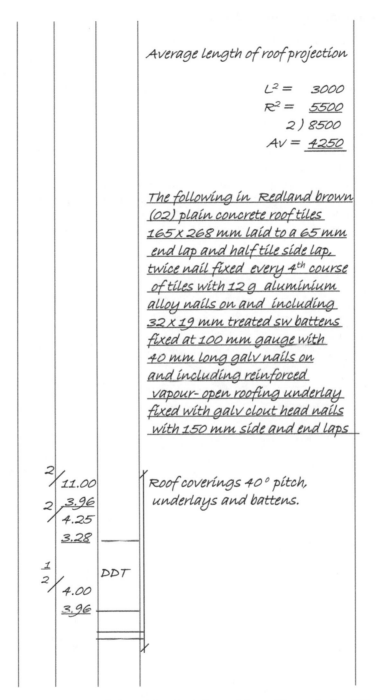

Average length of roof projection

$$L^2 = \quad 3000$$
$$R^2 = \quad \underline{5500}$$
$$2\,)\,8500$$
$$AV = \quad \underline{4250}$$

The following in Redland brown
(02) plain concrete roof tiles
165 x 268 mm laid to a 65 mm
end lap and half tile side lap,
twice nail fixed every 4th course
of tiles with 12 g aluminium
alloy nails on and including
32 x 19 mm treated sw battens
fixed at 100 mm gauge with
40 mm long galv nails on
and including reinforced
vapour- open roofing underlay
fixed with galv clout head nails
with 150 mm side and end laps

$2/\ 11.00$

$2/\ \underline{3.96}$
$/\ 4.25$
$\underline{3.28}$

Roof coverings 40° pitch,
underlays and battens.

$\frac{1}{2}$
$2/\ 4.00$
$\underline{3.96}$

DDT

Figure 10.5b An example of booked dimensions where the roof slope of the projection differs from the main roof slope.

10.02.02 Descriptions for pitched roof coverings

Interlocking concrete tiles are the most commonly used roof coverings since they are relatively cheap to purchase and fix. They are manufactured in a number of different profiles, colours and finishes (see Figure 10.6). The large-scale use of the more traditional roofing materials such as clay plain tiles and natural slates is limited by the cost of supply and fixing (see Figure 10.7). Both are manufactured in the form of cheaper alternatives (concrete plain tiles and fibre cement slates). Even so, they are still a more expensive roof covering than an interlocking concrete tile. Assuming a standard pitched roof, once the roof or wall covering area has been booked (square metres), the pitch of the roof is given in the description. This is followed by the measurement of boundary work, given in linear metres, with a dimensioned description stating the net girth of the type of boundary work involved. Boundary items include abutments, eaves, ridges, verges, valleys, hips and vertical angles. In addition, it will be necessary to include in the description the incline or plane for each of these boundary items as follows: horizontal, sloping, raking, vertical, curved (radius stated), stepped and preformed.

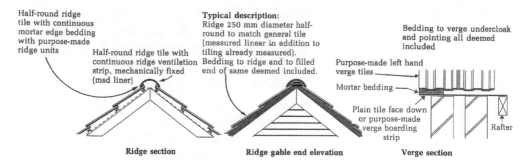

Figure 10.6 Interlocking concrete tiling.

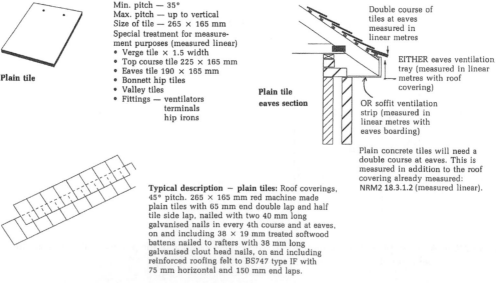

Figure 10.7 Plain roof tiling.

This is followed by the measurement of pitched roof fittings (recorded as enumerated items) and described as a dimensioned description (or manufacturer catalogue ref.). This would include ventilators, finials, gas terminals, hip irons, soakers, saddles, rooflights and the like.

Mandatory information provided should include the type, quality and size of materials:

- type of tile, slate or shingle;
- spacing and type of battens and counter-battens;
- underlay type, laps and method of fixing;
- minimum extent of side and end laps;
- method of fixing and number of fixings;
- composition and mix of mortar;
- type of pointing.

10.03 Flat roof coverings

The rules for the measurement of asphalt and built-up felt roof coverings are detailed in NRM2 17 Sheet roof coverings and NRM2 19 Waterproofing. Sheet metal roof coverings are included in NRM2 17. Measurement of flat roof coverings should follow the same pattern as pitched roof coverings, with the roof structure being measured first, followed by the covering.

10.03.01 Mastic asphalt roofing (NRM2 17)

Asphalt roofing is measured in square metres, giving the area in contact with the base. The roof pitch is given in the description and the width must be classified as either >500 mm or <500 mm. The former is booked in square metres while the latter is recorded in linear metres. Skirtings, fascias, aprons, gutter linings, channels, valleys and kerbs are measured in linear metres, stating the girth in the description in terms of the respective net girth on face together with the plane or angle of installation: horizontal, sloping, vertical, raking, curved (stating the radius) or stepped. As with associated work sections, there is a long list of items that are deemed included (internal angle fillets, fair and rounded edges, drips, arises and turning asphalt nibs into grooves are all deemed included). NRM2 19.10.1.1–7 allows for the measurement of 'spot items' as enumerated items and would include catch pits, sumps, outlets and the like. Any fittings such as ventilators are also measured as enumerated items (NRM2 19.11.1.1/2). Edge trim and preformed angle trim are measured in linear metres and are deemed to include ends, angles and intersections (NRM2 19.12).

10.03.02 Built-up felt roofing

This category is measured in the same fashion as asphalt roofing, in square metres. The area recorded in the dimension column is the area of felt in contact with the base, stating in the description the pitch of the roof. Details of the quality and kind of felt, together with the nature of the base, the method of jointing and the height of the work above ground level, are all required to be given in the description. Where the width of the roof covering is greater than 500 mm, the covering is measured in square metres; where it is less than or equal to 500 mm wide, it is measured in linear metres (NRM2 17.1 and 17.2). The description should include reference to the roof slope/plane as follows: horizontal, sloping (pitch stated) or vertical. Where curved work is involved, this must be stated together with the radii (NRM2 17.1/2.1–4). Any underlay, insulation and the finish of the exposed surface must also be described, together with the nature of the base that is being covered. A roof finish of solar reflective paint, stone chippings or the like should also be included here. Where an alternative roof finish is required – for example, tiles, paving slabs, grass roofs – these should be measured using the rules for an alternative work section. Associated labour items – drips, welts, rolls, seams and laps – are all measured as 'extra-over' the initial roof area in linear metres, stating in the description the respective dimensions (NRM2 17.3.1–5). Boundary work to voids is only measured where the void exceeds 1.00 square metres. The approach to the measurement of boundary work is consistent with the interpretation adopted for NRM2 work sections 18 and 19.

10.03.03 Sheet metal roof coverings

Lead, aluminium, copper, stainless steel and zinc sheet coverings are measured in square metres, stating the pitch of the roof in accordance with NRM2 17. This is the same set of rules used for measuring felt roofing. In the same fashion, drips, welts, rolls, seams and laps are all measured 'extra-over' the initial roof area. Adjustments for voids are only made where the void area exceeds one square metre. Boundary work caused by the inclusion of voids is measured only where the void area exceeds one square metre.

10.03.04 Sheet metal flashings/weathering

It is often necessary to measure metal flashings as part of the roof covering, to ensure a weather-tight construction. Principally this occurs where the slope of the roof is interrupted by chimney stacks or abutting walls. Traditionally the material used for this purpose is lead, which is available in a number of thicknesses referred to by codes (BS EN 12588). Code 4 (1.8 mm) is adequate for most situations, whilst code 5 (2.24 mm) would be used in conditions of severe exposure or for long lengths.

NRM2 17.5 provides the rules for measuring flashings and follows the rules for measuring sheet roof coverings. These are given in linear metres and are deemed to include undercloaks; rough and fair cutting; bedding; pointing of ends, angles and intersections; welted, beaded or shaped edges; and all dressing. The description should give the net girth and the type of flashing, be it a flashing, an apron, weathering, capping or kerbs (for a full list of options, see NRM2 17.5.1.1–9). Finally, the description should define the plane of the flashing: horizontal, sloping, vertical, raking, curved (radius stated) stepped or preformed.

Gutters and valleys (NRM2 17.6 and 17.7) are also measured using a similar approach (i.e. in linear metres, stating the net girth in the description, together with the plane or type of work including whether it is sloping, stepped, curved (radius stated), secret or tapered). Mention should also be made as to the nature of the base, the spacing of any structural supports and whether the work is preformed (NRM2 17.6.1.1–5.1–3 and NRM2 17.7.1.1–5.1–3). Finally, spot items and fittings are enumerated. Examples of spot items (NRM2 17.8.1–7) include catchpits, sumps, outlets, collars or sleeves around pipes, canopies and hatch covers. Fittings (NRM2 17.9.1–8) are deemed to include joints, dressing and bonding to the surrounding work and would include ventilators, finials, gas terminals, hip irons, soakers, saddles and roof lights. Where the method of fixing is not at the discretion of the contractor, this will need to be stated. As an alternative to a dimensioned description, a proprietary reference can be given.

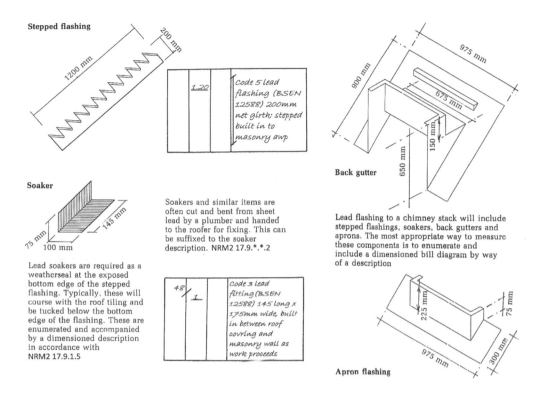

Figure 10.8 Chimney stack leadwork.

Typically, the leadwork to a chimney stack will require the fabrication of a number of metal components including front apron, back gutter, stepped flashings and soakers. These components, together with typical measured items for each, are shown in Figure 10.8.

10.4 Fascia, eaves and verge boarding (NRM2 16.3, NRM2 16.4)

Eaves soffit boarding (fascia and soffit), together with verge (barge) boards not exceeding 600 mm girth, are measured in linear metres giving their size in the description (NRM2 16.4.1). Where boards exceed 600 mm in width, they may be measured superficially (NRM2 16.4.2). Individual or framed supports for eaves soffit boards, together with decorations, can be measured at the same time. Simple grounds and battens are measured in linear metres (NRM2 16.3.1.1/2), while framed grounds, battens and bracketing should be given in square metres, stating the centres of the members (NRM2 16.3.2.1.1). In each case the nominal dimensions of the timber members should be given in the description. Boarding to eaves, fascia and bargeboards should be measured in linear metres where this work does not exceed 600 mm in width (NRM2 16.4.1.5); and in square metres where it exceeds 600 mm in width (NRM2 16.4.2.5). This should be described as work to soffits, and state the finish required where it is not to be left as sawn. There is also a requirement to state the location of work such as eaves, verges and the like (Figure 10.9; NRM2 16.4 notes, comments and glossary).

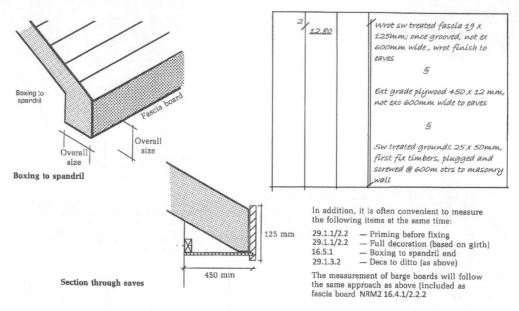

Boxing to spandril

Boxing to spandril

Fascia board

Overall size

Overall size

Section through eaves

125 mm

450 mm

2 / 12.80   *Wrot sw treated fascia 19 x 125mm; once grooved, not ex 600mm wide, wrot finish to eaves*

&

Ext grade plywood 450 x 12 mm, not exc 600mm wide to eaves

&

Sw treated grounds 25 x 50mm, first fix timbers, plugged and screwed @ 600m ctrs to masonry wall

In addition, it is often convenient to measure the following items at the same time:

29.1.1/2.2   — Priming before fixing
29.1.1/2.2   — Full decoration (based on girth)
16.5.1   — Boxing to spandril end
29.1.3.2   — Decs to ditto (as above)

The measurement of barge boards will follow the same approach as above (included as fascia board NRM2 16.4.1/2.2.2

Figure 10.9 Fascia, eaves and verge boarding.

Eaves to gable end roofs require boxing at exposed ends (see Figure 10.9, spandril end). These can be grouped with ornamental ends of timber members and are measured as enumerated items, stating in the description the kind, quality and finish of the material together with the overall size (NRM2 16.5.1). Depending on the construction, the eaves soffit may include continuous ventilation. Whilst no specific mention is offered in NRM2 for eaves ventilation, it is suggested that any vent units are measured at the same time as the eaves boarding using NRM2 16.6.1.2.8 as a general guide.

Two separate decoration operations are normally required for all timber surfaces. The first is a priming or sealing application, which is applied prior to the fixing or assembly of any joinery/timber components. It is normal to knot and stop timber prior to the application of primer; a single description can be written to include all three of these operations (knotting, priming and stopping). The second operation is carried out once the timber item has been assembled, and combines the separate application of three further coats of paint (two undercoats and one finishing coat) in a single description.

The rules for measuring paintwork distinguish between work under 300 mm in width (NRM2 29.1–7.1), which is measured in linear metres, and work in widths exceeding 300 mm (NRM2 29.1–7.2), which is measured in square metres. Before dimensions can be recorded, it will be necessary to identify which unit of measurement applies and a waste calculation should be prepared to check the decorating girth. This is illustrated in Figures 10.10. In this particular instance, it is necessary to describe any painting to eaves as 'external'. In addition, it is advisable to check the height of the eaves or barge boarding above floor or ground level. Where this exceeds 3.50 m, this should be stated in the description in accord with NRM2 29.1.1–3.2.1. Further details and examples of measuring decoration items can be found in Chapter 12, Internal finishes.

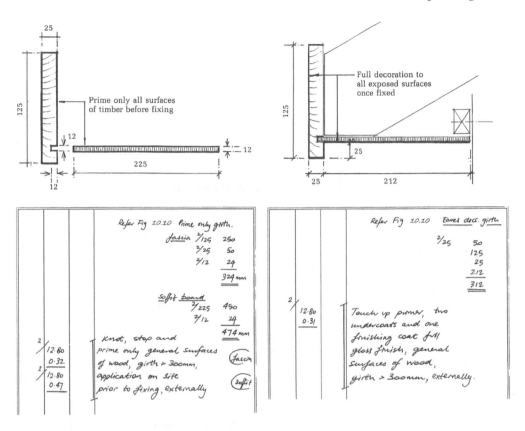

Figure 10.10 Measurement of paintwork.

10.05 Rainwater gutters and downpipes

The measurement of rainwater goods completes the work associated with roof coverings. Gutters and downpipes are measured in linear metres over all fittings (NRM2 33.1.1.1–3.1.1 and NRM2 5.1.1–3.1.1). The description should include reference to the type of pipe or gutter and the nominal size, together with details of the method of fixing and the type of background. Gutter and pipe brackets, together with running joints/couplings, are deemed included. Pipework and gutter ancillaries, such as running outlets, stopped ends, offsets and connecting shoes, are enumerated (NRM2 33.2.1 and NRM2 33.5.1). Downpipe fittings should be described as either less than or equal to 65 mm diameter or exceeding 65 mm diameter. In either case the description should include the number of connections made to the pipework in accordance with NRM2 33.3.1/2.1–4. Plastic (uPVC) rainwater goods are manufactured in different colours and will not normally require any decoration. Where metal gutters and downpipes are specified, it is likely that these will require decoration and this should be carried out in accordance with NRM2 29.5 for gutters and NRM2 29.6 for pipework (Figure 10.11).

A sample take-off for a traditional cut pitched roof with plain tile roof covering follows.

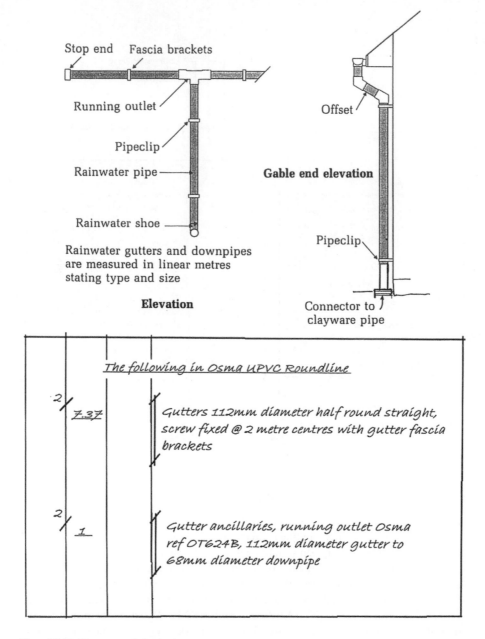

Figure 10.11 Gutters and downpipes.

10.06 Plain tiled pitched roof drawing and specification (see also p. 220)

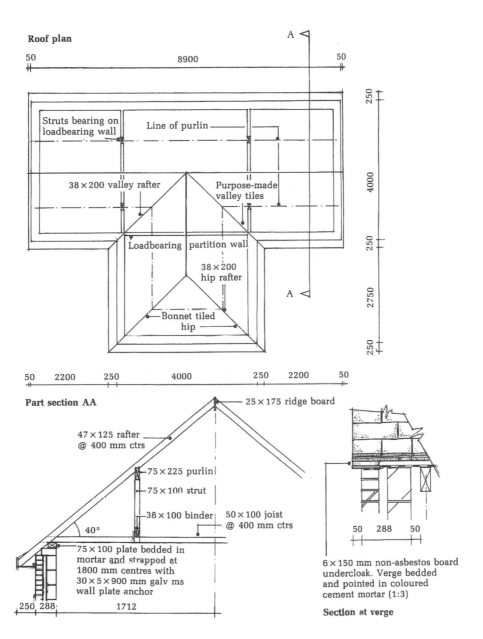

Roof plan

50   8900   50

A

250
4000
250
2750
250

Struts bearing on loadbearing wall

Line of purlin

38 × 200 valley rafter

Purpose-made valley tiles

Loadbearing partition wall

38 × 200 hip rafter

A

Bonnet tiled hip

50   2200   250   4000   250   2200   50

Part section AA

25 × 175 ridge board

47 × 125 rafter @ 400 mm ctrs

75 × 225 purlin

75 × 100 strut

38 × 100 binder

50 × 100 joist @ 400 mm ctrs

40°

75 × 100 plate bedded in mortar and strapped at 1800 mm centres with 30 × 5 × 900 mm galv ms wall plate anchor

250 288   1712

50   288   50

6 × 150 mm non-asbestos board undercloak. Verge bedded and pointed in coloured cement mortar (1:3)

Section at verge

(See p. 220 for section details at eaves and ridges together with specification marks.)

Detail section at eaves

268 × 165 mm plain concrete tiles
laid to a 100 mm gauge on 32 × 19 mm
preserved sw battens on reinforced felt
type 1F to BS 747

150 mm thick glass fibre insulation

Eaves ventilator unit fixed
to feet of rafters

Eaves vent fascia
grille

100 mm dia. half
round plastic gutter

150 × 25 once grooved treated
sw fascia

6 mm thick ext. quality ply
soffit board on 38 × 38 sw
treated ground plugged and
screwed to brickwork

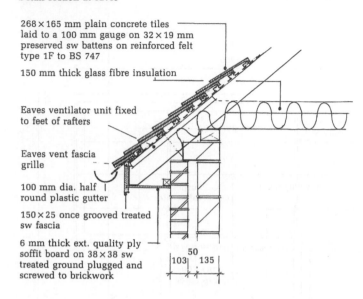

|50|
|103|135|

Half round ridge tile 450 mm long
mechanically fixed with stainless
steel nails to 50 × 50 mm treated
sw ridge batten

Continuous
ridge ventilating
strip

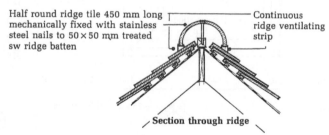

Section through ridge

Specification notes

Structural and non-structural timbers pressure impregnated with
  preservative.

All structural timber stress graded to CP 112 grade SS group S1.

Continuous ridge and eaves ventilation using dry vent ridge system and
  plastic eaves ventilation tray all to BS5250.

Base coat and two finishing coats wood stain to eaves.

UPVC gutters and downpipes to BS4576.

Redland plain concrete tiled roof (brown 02) with purpose made valley tiles
  and bonnet tiled hips.

10.07 Measurement example (Job ref: BMMS 7476/HH)

Pitched roof with plain tiled roof covering   Job ref: BMMS 7476/HH   Page 01 of 09

The work comprises plain tile roof covering to a pitched timber roof for a residential property, all in accordance with roof plans, sections and details provided

Take off list

TRADITIONAL PITCHED ROOF WITH PLAIN TILE ROOF COVERING

NRM2 requires the following drawings to be available:-
Roof plan/s, principle roof sections, external elevations.

Mandatory information to be provided for carpentry and roof coverings :-
See NRM2 work sections 16 and 18 for mandatory information to be provided .

Take-off list

ROOF COVERINGS
- Roof tiling
- Boundary work
  - Eaves and eaves ventilation
  - Verges
  - Ridge and ridge ventilation
  - Hips
  - Valleys

EAVES
- First fix eaves timbers (grounds)
- Fascia and soffit boarding
- Decoration to fascia and soffit (external work)
- Spandril ends

RAINWATER GOODS
- Guttering
- Gutter ancillaries
- Downpipes
- Downpipe ancillaries

Pitched roof with plain tiled roof covering   Job ref: BMMS 7476/HH   Page 02 of 09

The work comprises a plain concrete tiled roof for a two storey residential property, all in accordance with floor plans, sections and specification details provided

The following in Redland brown
(02) plain concrete roof tiles
165 x 268 mm laid to a 65 mm
end lap and half tile side lap,
twice nail fixed every 4th course
of tiles with 12 g aluminium
alloy nails on and including
32 x 19 mm treated sw battens
fixed at 100 mm gauge with
40 mm long galv nails on
and including reinforced
vapour- open roofing underlay
fixed with galv clout head nails
with 150 mm side and end laps

Roof coverings
Main roof length   8 900
add verge o'hang 2/50   100
                             9 000

Projectg roof length   2 750
add eaves o'hang   250
                             3 000

Main roof and projecting roof slope
length

rafter length (from CH09) 3 036
add
eaves tile projection
            into gutter   25
                             3 061

Roof coverings (Contd)

2/	9.00	
	3.06	
2/	3.00	
	3.06	

Roof coverings 40° pitch, underlays and battens.

NRM2 18.1.1.1

Eaves

2/	9.00	
2/	3.00	

Boundary work 165 x 200 mm eaves tile each tile twice nail fixed, sloping

NRM2 18.3.1.2.2

§

Fittings Redland plastic fluted eaves ventilation units, nail fixed to timber feet of rafters

NRM2 18.4.1.1.1

Verges

2/		
2/	3.06	

Boundary work, to gable ends, verges, to sloping roof including undercloak bedded and pointed in c.m. (1:3) each tile twice nailed with 12 gauge aluminium nails

NRM2 18.3.1.4.2.

Ridge

9.00	
3.00	

Boundary work, to apex of roof, half round pre-holed ridge tile fixed with 2 nr annular shanked stainless steel nails

NRM2 18.3.1.3.2.

§

Fittings Redland plastic Uni-Vent Rapid Ridge/Hip, ventilators, screw fixed to timber ridge board.

NRM2 18.4.1.1.1/2

Pitched roof with plain tiled roof covering   Job ref: BMMS 7476/HH   Page 04 of 09

<u>Roof coverings</u> (Contd)

<u>The following in Redland brown</u>
<u>(02) plain concrete roof tiles</u>
<u>165 x 268 mm (Contd)</u>

<u>Hip and valley rafter length</u>
Half span (as before)   2 250
<u>add</u>
eaves tile proj
        into gutter   <u> 25</u>
                    <u>2 275</u>
slope tiled
   length (as before)   <u>3 036</u>
Slope calc using Pythagoras
Length =
$\sqrt{\text{half span}^2 + \text{slope length}^2}$

$\sqrt{2.275^2 + 3.036^2}$

   = <u>3. 793</u> slope length of
              hip and valley

2/ 3.81   Boundary work , valleys in   (valleys)   NRM2 18.3.1.5.2
       Redland concrete valley tile
       nail fixed to timber rafters
       sloping

                     &

       Boundary work, bonnet hip   (hips)
       tiles, 325 mm wide nail fixed
       with aluminium alloy nails   NRM2 18.3.1.6.2
       and bedded in c.m. (1:3),
       sloping.

Roof coverings (Contd)

Eaves boarding – fascia

main roof		8 900
add	verges 2/50	100
		9 000

proj hipped roof		2 750
add	eaves proj 2/250	500
		3 250

fascia girth	2/150	300
	2/25	50
	2/12	24
		374

2/
2/ 9.00
3.25

Backing and other first fix timbers, 38 x 38mm treated sw grounds

§

NRM2 16.3.1

Sw boarding, fascias and the like not exceeding 600mm wide x 25mm thick, wrot finish

§

NRM2 16.4.1.1.1

Painting (base coat staining) to general surfaces > 300mm girth, external, application on site prior to fixing

NRM2 29.1.2.2.6

x 0.37 = _____ m²

Pitched roof with plain tiled roof covering   Job ref: BMMS 7476/HH   Page 06 of 09

<u>Eaves boarding</u> (Contd)
        <u>- soffit</u>

<u>Length of soffit</u>
Main gable roof   8 900
Verge proj 2/50   100
                         9 000

  ²/9 000   18 000
Proj hip roof
         ²/   2 750
eaves /250   500
         ²/ 3 250   6 500
<u>soffit length</u>   <u>24 500</u>

Soffit width   250
<u>less</u> fascia   25
<u>add</u> groove   12   <u>13</u>
                           <u>237</u>

<u>Soffit girth</u>
          2/237   474
          2/6   <u>12</u>
                       <u>486</u>

24.50	Plywood boarding, soffits and the like not exceeding 600mm wide x 6 mm thick, horizontal, wrot finish	NRM2 16.4.1.1.1
	&	
	Painting (base coat staining) to general surfaces > 300mm girth , external, application on site prior to fixing	NRM2 29.1.2.2.6
	x 0.49 = m²	

Eaves boarding (Contd)
      - spandril end

2/ 2	Ornamental ends of timber members, 19mm thick external quality plywood spandril end to eaves overall size 250 x 450 mm	NRM2 18.3.1.5.2

&

| | Painting (base coat staining) to general surfaces, isolated areas < 1.00 m² , external, application on site prior to fixing | NRM2 18.3.1.6.2 |

Decorations to eaves
      Exposed girth of eaves
      eaves overhang   250
      fascia   150
      return 2/25   50
                            450

| 2/2/ | 24.50
0.45
0.25
0.43 | Painting, two coat stain application to general timber surfaces > 300mm girth, external, work to eaves exceeding 5.00 – 8.00 m above ground level | NRM2 29.1.2.2.1

Decoration to spandril ends included with with general eaves decoration therefore not considered an isolated area |

Gutters and Downpipes

main roof   8 900
add   verges 2/50   100
                       9 000

proj hipped roof   2 750
add eaves proj   250
                       3 000

To centre of gutter
      2/2/1/
          2/ 100   200
                     3 200

Pitched roof with plain tiled roof covering   Job ref: BMMS 7476/HH   Page 08 of 09

<u>Gutters and Downpipes</u>
(Contd)

2/		
2/	9.00	Gutters 100mm Ø half round uPVC, straight with combined fascia bracket and running connectors @ 2.00 metre centres screw fixed to sw fascia boards
	3.20	

NRM2 33.5.1.1.1

<u>The following in gutter ancillaries (spec as above)</u>

2		100mm to 68mm Ø running outlet
4		100mm Ø half round stopped end
2		100mm Ø half round external angles
2		100mm Ø half round internal angles

NRM2 33.6.1

<u>Assumed waste calculations for height of downpipe</u>

G floor to fist floor			2 400
GL to dpc			150
floor joists		225	
floor boardg		20	245
			2 795
first floor to eaves	2 400		
less eaves proj		350	2 050
			4 845

2/	4.95	Pipework 68mm Ø uPVC downpipe, straight with push fit socket joints, secured in pipe brackets plugged and screw fixed to masonry work @ 2.00 metre centres

NRM2 33.1.1.1.1

			Pitched roof with plain tiled roof covering Job ref: BMMS 7476/HH Page 09 of 09	
			Gutters and Downpipes (Contd)	
	2		*The following in pipework* *ancillaries (spec as above)* 68mm ∅ for offset bend (swan neck) with 250mm projection	NRM2 33.1.1.1.1

11 Windows, doors and standard joinery

11.01 Introduction

In the previous edition of the Standard Method of Measurement (SMM7), windows, doors, timber staircases and glazing all shared a common work section. NRM2 adopts an approach that sees each of the above allocated an individual work section as follows: NRM2 Work Section 23 includes windows, screens and lights; Work Section 24 covers doors, shutters and hatches; and Work Section 25 is intended for timber stairs, walkways and balustrades. Glazing work that is not supplied as part of a window or door unit is measured under a bespoke Glazing work section (NRM2 Work Section 27). Associated architraves, window boards and sealant joints are included as part of General Joinery (NRM2 Work Section 22). Typically these associated General Joinery items would not form part of any standard joinery component so they are grouped, for measurement purposes, along with skirtings, picture rails, isolated shelves, worktops and isolated handrails. They all share a common set of measurement rules and unit of measurement (linear metres).

The separate treatment of windows, doors and staircases will only be evident in the finished Bill of Quantities. For measurement purposes, most of these joinery items will form an integral part of the measurement of a larger component (e.g. architraves with doors, window boards with windows and isolated handrails with staircases).

11.02 Scheduling windows and doors

As with the measurement of surface finishes, there is a degree of repetition in the measurement of windows and doors. In order to maintain a consistent and efficient approach, the preparation of a schedule to assist in the measurement of both windows and doors is advised. NRM2 requires that a window and/or door schedule must accompany the documents that are provided with the measured work for these two work sections. In many cases, window and door schedules will be provided by the design team. Where window and door schedules are not available, it is advised that these are prepared before the process of booking dimensions commences. In the first instance (and assuming that the drawings have not already provided ID labels for the windows and doors), the floor plans should be marked to provide each window and external door with a unique reference. It is convenient to include external doors and windows in the same schedule. A separate schedule should be prepared before measuring internal doors. For most projects of any scale, time spent preparing window and door schedules speeds up the measurement process by allowing the easy identification of identical items, thereby facilitating measurement by avoiding unnecessary repetition when taking off (Figure 11.1).

11.03 Measuring windows and doors

When dimensions were recorded for the measurement of external walling (Chapter 6) and wall finishes (Chapter 12), no adjustment was made for window and external door openings. A note was made to this effect in the appropriate chapters. This may seem strange to the novice measurer, but is considered a standard measurement procedure as it facilitates a speedy and efficient take-off. It is not until this stage (measurement of windows and doors) that an adjustment will be made for this previous, deliberate over-measurement. Having first booked dimensions for the appropriate items for the window or external door unit/s, together with any glazing, decoration and ironmongery, the area of external or internal walling and associated wall finishings is adjusted. This is followed by the measurement of the lintel, closing cavities, damp proof courses, external and internal sills, together with plasterwork to reveal and soffits and any angle beads. The same approach is adopted for the measurement of

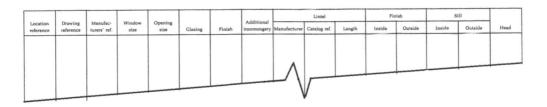

| Location reference | Drawing reference | Manufac- turers' ref. | Window size | Opening size | Glazing | Finish | Additional ironmongery | Lintel | | | Finish | | Sill | | Head |
| | | | | | | | | Manufacturer | Catalog ref. | Length | Inside | Outside | Inside | Outside | |

Figure 11.1 Window schedule headings.

rooflights, although the adjustment will only include the omission of roof coverings where the opening area exceeds 1.00 m² (NRM2 18.1/2 notes comments and glossary item 2). Roof members will require trimming, and there may be some additional measurement for boundary work (NRM2 18.3).

In the majority of cases, both windows and external doors are likely to be delivered to site as complete units ready for direct inclusion in the works. While windows and internal doors will be supplied complete with the frame, external doors are likely to require that the frame be provided as a separate item. This is recognised in NRM2 by the inclusion of a separate measured item for the supply of door frames (NRM2 24.9). Reference should be made to NRM2 24.1 for the measurement of external doors and frames supplied as composite items. A similar distinction is made for internal doors that will either be supplied as door sets (NRM2 24.1) or as two separate components – a door unit and a lining set (NRM2 24.2 and NRM2 24.10). Measurement in each case is relatively straightforward, with doors being measured as enumerated items (nr) giving a dimensioned description (NRM2 24.2.1), and door frames or door linings measured in linear metres supported by a cross-sectional dimensioned description (NRM2 24.9 and 24.10).

Mandatory information needs to be provided for door sets, doors with linings or frames and windows. This should include consideration of the items listed in NRM2 Work Sections 23 and 24. In most cases a manufacturer's name, the product range and a product reference, together with the size of the unit, will provide the estimator with sufficient detail for pricing purposes. Typically, where prefabricated components are used, the measurement of a standard window unit would require the following description (Figure 11.2).

Glazing, painting and pointing frames with mastic will each require separate measurement.

11.03.01 Glazing

Glazing panes and sealed glazed units are measured as enumerated items. A full description must be given, including the kind, quality and thickness of glass or sealed glazed unit, the method of glazing, together with the details of any glazing gaskets or glazing compound, the method of securing the glass and the nature of the frame or surround (i.e. what the glazing is fixed to). Having made an allowance for glazing in rebates, the pane size of any window or door glass is recorded (width × height) as part of the written description (NRM2 27.1–3.2). The description would also need to include the thickness of the glass or overall thickness of any sealed unit (NRM2 27.1–3.1). Glazing in bathrooms or cloakrooms will require obscured glass, and there are additional rules for the measurement of any grinding, sandblasting, acid etching, embossing or engraving, and these are included under NRM2 27.4. 1–5. Specific rules are provided for the measurement of lead lights, mirrors, saddle bars and the removal and preparation of a frame to receive new glass. Each of these merits a specific set of measurement rules (NRM2 27.5 – 27.8).

11.03.02 Painting and decorating windows and doors

For measurement purposes, painting and decorating of items associated with windows and doors can be considered in two parts. The first is to the wall area occupied by the window or door opening (this would normally be an emulsion paint finish on two-coat wall plaster or plasterboard dry lining). The second is to the decoration of the door or window unit itself (this would normally be oil-based paint or varnish finish, assuming a timber window/door).

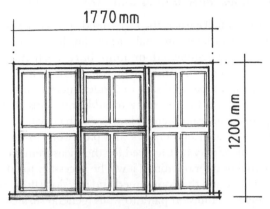

Ref:- Jeld-wen LEWFG312CDC
Window opening size 1700 x 1200mm
Window size 1695 x 1195mm

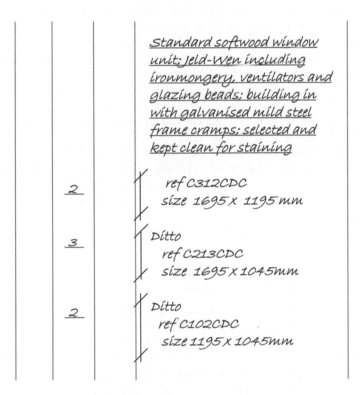

		Standard softwood window unit: Jeld-Wen including ironmongery, ventilators and glazing beads; building in with galvanised mild steel frame cramps; selected and kept clean for staining
2		ref C312CDC size 1695 x 1195 mm
3		Ditto ref C213CDC size 1695 x 1045mm
2		Ditto ref C102CDC size 1195 x 1045mm

Figure 11.2 Measurement of standard window units.

11.03.03 Forming openings for windows and external doors

Opening adjustment: as identified earlier, the recognised approach when booking dimensions for external walls and any associated internal surface finishes ignores all window and door openings. It is only once the window/door units have been measured that this over-measurement is adjusted. In addition to deducting the skins of a cavity wall (together with any cavity wall insulation), it will be necessary to deduct internal surface finishes (wall plaster and emulsion paint). It will also be necessary to add any items that were not previously included. This will require the measurement of lintels, closing cavities, cavity trays and vertical damp proof courses together with plasterwork and paintwork to reveals and soffits. In the specific case of windows, there may be the need for the measurement of an external sill and an internal window board. For door openings, it will be necessary to adjust (deduct) the length of the skirting board and to extend the floor screed and any floor finish into the door opening.

The initial measurement of emulsion paint to walls (general surfaces) also presumed gross measurement, i.e. it ignored all window and door openings (see masonry, Chapter 6 and surface finishes, Chapter 12). It is only at this stage that 'an adjustment' to the decorated wall area for any window or door openings will be made. This may seem odd, but is recognised as an efficient approach to measurement since the opening dimensions of any window and door units are now readily to hand. The 'adjustment' is a simple deduction to the wall decoration (emulsion paint) area that has previously been measured. The area of this deduction will be the elevation area of the window or door being measured, and will be based on the length and height of the window/door in question. The resultant quantity of wall decoration area will not be evident until these two separate measurements are brought together at the abstracting stage. It will also be necessary to include an additional measurement for emulsion paintwork to the reveals and soffits of any windows or doors. One way to envisage this would be to imagine an opening being cut with an angle grinder on the internal elevation of a plastered and decorated cavity wall. Because of the depth of the cavity, it means that both plasterwork and decoration will be necessary to the now exposed sides (reveals) and to the underside of the top (soffit) of any opening. At the same time and using the same adjustment area, a deduction will be made for the previous deliberate over-measurement of the external walling. This would typically include internal blockwork, forming the cavity, cavity wall insulation and external brickwork.

It may be helpful at this point to review the measurement of masonry work (Chapter 6) and surface finishes (Chapter 12).

11.03.04 Decorating windows and doors

In regard to any decoration to window units or doors, NRM2 Work Section 29 provides two classes of work that may be applicable: NRM2 29.1 caters for blank doors and NRM2 29.2 for glazed doors and windows. It is assumed that in most cases paintwork to windows and doors will fall into the level one description, 'exceeding 300 mm girth' (NRM29 1/2.2), and, as a consequence, will require measuring in square metres. It should also be noted that internal and external paintwork has to be appropriately identified and described separately (NRM2 29.1/2.2.1 or 29.1/2.2).

Whether measuring paintwork to windows, part or fully glazed doors or door blanks, the same approach can be adopted. The elevation area of the window is recorded, and is measured over the glazed area including the frame, together with any transoms and mullions. Unlike previous editions of the Standard Method of Measurement the size of the glazed area (pane area) can be ignored, as is evident from the description 'painting to glazed surfaces irrespective of pane size' (NRM2 29.2). The paintwork area should be measured to each face of a window and should be calculated to include work to edges of opening lights. As mentioned earlier, typically this will exceed 300 mm in girth and will be measured in square metres. Work to glazed doors is classified in a similar fashion, and should also include an allowance for the opening edges of doors. Paintwork to door blanks follows a similar approach, but would be described as 'painting to general surfaces' (Figure 11.3; NRM2 29.1).

Standard timber window

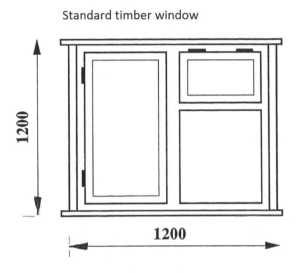

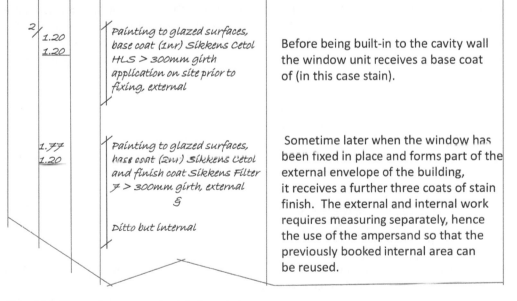

Figure 11.3 Decoration to a standard timber window.

11.03.05 Bedding and pointing window and door frames

Assuming traditional cavity wall construction, windows and external doors will require building-in or subsequently securing to external walls. Traditionally this has been achieved by a galvanised metal fixing (wall tie) which is bent through 90 degrees, with one end screw-fixed to the vertical part of the timber frame and the other built into the mortar joint as the wall is constructed. Alternatively, window and door openings will be formed as the external walls are built, so that window and external door units can be screw-fixed into the formed opening. NRM2 assumes that the bedding and pointing of window and door frames is 'deemed included' (NRM Work Sections 23 and 24 'Works and materials deemed included'), and consequently requires no further mention. Where non-standard or specialist approaches are specified, this must be included as a separate linear item in accordance with NRM2 14.21, stating the type of filler, sealant and pointing together with the method of application and any preparation. Where both are required, it is suggested that the two operations are combined in a single description.

For reference purposes, an extract from a typical Bill of Quantities (BQ) is included showing a number of standard windows grouped under an 'all-embracing' Level 3 description. Both pointing frames with mastic and glazing are included on the same BQ page (Figure 11.4).

11.04 Internal doors

Standard internal doors, like windows, should be enumerated and are best described by referring in the description to the type or style of door, the manufacturer's code and the door unit dimensions (height, width and door thickness). This general pattern replicates the two-stage approach adopted for the measurement of windows. Initially, the type of door, the door size, any ironmongery and surface finishes are measured, followed by an adjustment for any items required to form the opening. Where a number of internal doors are proposed and an internal door schedule is not provided as part of the design information, it is advisable to prepare one. With the exception of internal fire doors, door linings are more likely to be supplied as separate components to the internal door leaf and this will require the measurement of the lining as a separate item (NRM2 24.10). Door linings should be measured in linear metres along with any door stops (NRM2 24.11), giving the cross-sectional dimension of the lining or door stop in the description. Any associated fire stops and smoke stops are given in the same fashion (NRM2 24.12 and 24.13). While it is not a requirement of NRM2, it may be helpful for estimating purposes to give the number of identical door lining sets in the description (assuming a composite door lining set). By and large, most door lining sets are supplied in standard sizes to suit both commonly used internal door sizes and internal wall partition widths. Care should be exercised when establishing the door lining width to ensure that, in addition to the partition width, the dimension includes the thickness of any plasterwork/plasterboard (or other finishing) to both sides of the formed opening (Figure 11.5).

When not supplied as part of the door leaf, the provision of glazing and ironmongery, together with decorating to internal doors, should be measured using the same approach as previously described for windows and external doors.

Samuel Thomas Recycling Centre					£	p
23 WINDOWS, SCREENS & LIGHTS						
Windows and window frames;						
Standard softwood window unit; Jeld-Wen including ironmongery, ventilators and glazing beads; building in with galvanised mild steel frame cramps; selected and kept clean for staining						
A ref C312CDC; size 1695 x 1195 mm	3	nr				
B ref C213CDC; size 1695 x 1045mm	4	nr				
C ref C102CDC; size 1195 x 1045mm	3	nr				
Glazing						
Glass, sealed double glazing units comprising 4mm thick low emissivity glass (2nr) and 12mm gas filled airspace to timber frames and casements with non-setting mastic and wood beads,						
D pane size 500 x 250mm, glazing rebates 20 -30mm	24	nr				
E pane size 450 x 250mm, glazing rebates 20 -30mm	48	nr				
F pane size 300 x 250mm, glazing rebates 20 -30mm	30	nr				
Page 3/87			To Collection			

Figure 11.4 Sample NRM2 Bill of Quantity page presentation for window units and associated glazing in the Bill of Quantities.

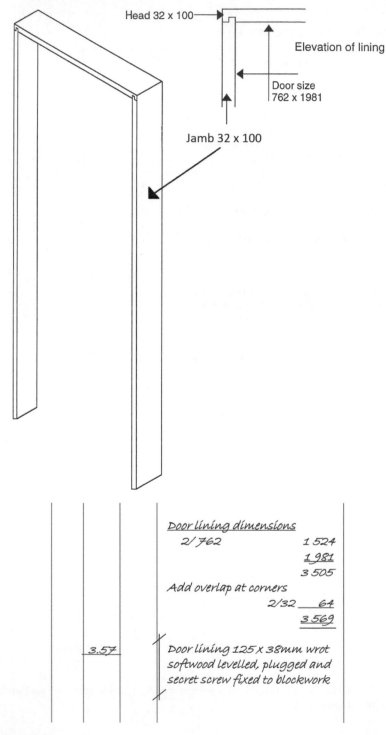

Figure 11.5 Door lining.

11.5 Timber staircases

As with windows and doors, the majority of timber staircases are prefabricated and delivered to site ready to be included in the works. NRM2 includes a specific work section for the measurement of timber, metal and plastic staircases together with associated balustrading, newel posts and handrails (NRM2 25). A range of standard staircase components is available, from which it is possible to prefabricate a variety of different staircase configurations. These are delivered to site prefabricated, and are conveniently described using the manufacturer's coding system, recorded as enumerated items in accordance with NRM2 25.1.1/2. The inclusion of quarter- and half-landings is enumerated and described as 'extra-over' the staircase on which they occur (NRM2 25.4). Attached balustrades and newel posts are deemed included. If either balustrades or newel posts do not form an integral part of the staircase, they should be measured separately – either enumerated or in linear metres (Figure 11.6; NRM2 25.7–11).

As part of the opening adjustment, it may be necessary to adjust any previous measurement of intermediate floor joists and floor boarding. At the very least, it will be necessary to check to ensure that the stairwell opening has been formed and, if not, the necessary trimming adjustment should be made (for forming of structural openings in timber floors, see Chapter 9). An equivalent adjustment (deduction of previous over-measurement) is likely to be necessary for surface finishing ceiling items plasterboard with plaster finish and decoration.

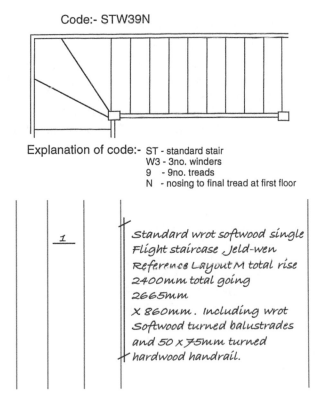

Figure 11.6 Standard stairs.

11.06 Purpose-made joinery

There is no specific mention for the measurement of purpose-made windows, doors or staircases in NRM2. However, there is sufficient flexibility in NRM2 work sections 23, 24 and 25 to allow the measurer to provide sufficient details for pricing bespoke joinery items. For non-standard windows, doors and staircases, NRM2 23.1, 24.2 and 25.1, these should be enumerated with a dimensioned description or diagram. Alternatively, see NRM2 Work Section 2; Off-site manufactured, materials, components or buildings. In the former instance, this would require a full description identifying the materials together with the various components, their respective sizes and the method of jointing or form of construction. There is a presumption throughout NRM2 that drawings, schedules and general arrangement plans will accompany the tender documentation. Where non-standard joinery items are specified and a component detail drawing is available, the obvious approach would be to enumerate the item in question and make reference to this in the description, ensuring that any drawing is included as part of the tender documentation. The latter approach is based on the assumption that component drawings will be available at bill production stage, and is by far the most convenient from a measurement point of view. Alternatively, where the design of the component is incomplete, the work should be the subject of a Prime Cost Sum in accordance with NRM2 Part 3, Tabulated Rules of Measurement for Building Works 3.3.7.

11.07 Skirtings, architraves and window boards

A number of other isolated joinery components, which would normally be measured with doors and windows, are included in NRM2 Work Section 22, General Joinery. The un-framed isolated trim items covered here are skirtings, architraves (NRM2 22.1 and 22.2) and window boards (NRM2 22.5). In all cases, the unit of measurement is linear metres stating in the description the dimensioned overall cross-section size. Ends, angles, mitres and intersections are deemed included, regardless of the cross-sectional area of the timber (Figure 11.7).

11.8 Ironmongery

Where not supplied as part of the prefabricated component, ironmongery is enumerated stating in the description the type, the manufacturer's reference, the method of fixing and the background to which it is fixed (NRM2 23.10). Most standard windows are supplied complete with all ironmongery, and the measurement of butts, latches, locks and handles will only be required for doors (NRM2 24.16). Where no decision has been made regarding the type or style of ironmongery, a Prime Cost Sum can be included (in accordance with NRM2 Part 3, Tabulated Rules of Measurement for Building Works 3.3.7.). Where this is the case, the measurer need only record booked dimensions for fixing ironmongery. These should be enumerated, stating in the description the type of ironmongery and the nature of the base to which the door furniture will be fixed. An internal door would typically require the ironmongery shown in Figure 11.8.

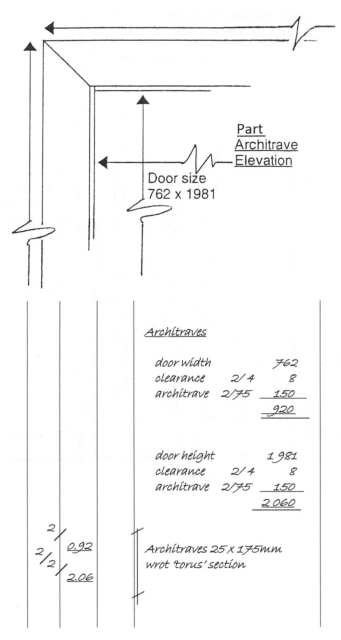

Figure 11.7 Internal door architrave (part elevation).

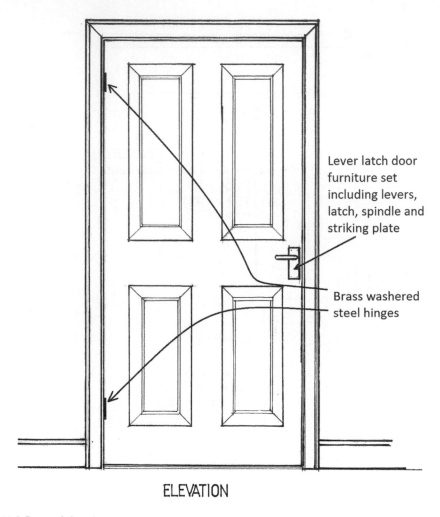

Lever latch door furniture set including levers, latch, spindle and striking plate

Brass washered steel hinges

ELEVATION

Figure 11.8 Internal door ironmongery.

11.9 Standard window drawing and specification

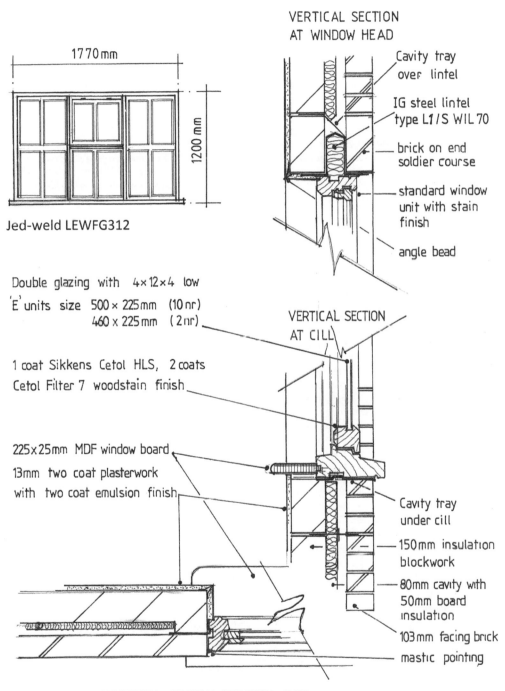

1770 mm

1200 mm

Jed-weld LEWFG312

VERTICAL SECTION
AT WINDOW HEAD

Cavity tray
over lintel

IG steel lintel
type L1/S WIL 70

brick on end
soldier course

standard window
unit with stain
finish

angle bead

Double glazing with 4×12×4 low
'E' units size 500 × 225mm (10 nr)
460 × 225 mm (2 nr)

1 coat Sikkens Cetol HLS, 2 coats
Cetol Filter 7 woodstain finish

225 × 25mm MDF window board

13mm two coat plasterwork
with two coat emulsion finish

VERTICAL SECTION
AT CILL

Cavity tray
under cill

150mm insulation
blockwork

80mm cavity with
50mm board
insulation

103mm facing brick

mastic pointing

HORIZONTAL SECTION THROUGH JAMB

11.10 Standard window measurement example (Job ref: ALJS 8587/Wdws)

Standard window unit   Job ref: ALJS 8587/Wdws   Page 01 of 07

The work comprises the provision and installation of standard window units for a residential property, all in accordance with floor plans, elevations and schedules provided

Take off list

NRM2 requires the following drawings to be available:-
General arrangement plans, Window, Glazing and Ironmongery Schedules.

Mandatory information to be provided for carpentry and roof coverings :-
See NRM2 work section 23 for mandatory information to be provided .

Take-off list

- Window and window frame
- Glazing
- Stain decoration
  - Internally
  - Externally
- Opening Adjustment
- Cavity wall
  - External skin
  - Cavity and Insulation
  - Internal Skin
  - Plasterwork
  - Decoration
- Lintel
- B-o-e soldier course
- Adjust bwk for ditto
- Cavity tray (dpc)
- Closing cavities at jambs
- Reveals/Soffits
  - Plasterwork
  - Decoration
- Dpc at sill
- Window Board
- Decoration to window board

Windows (Contd)

1	Standard softwood window unit; Jeld-Wen ref LEWFG312 size 1770 x 1200 mm including ironmongery, ventilators and glazing beads; building in with galvanised mild steel frame cramps; selected and kept clean for staining	NRM2 23.1.1.*.1 Manufacturer's Catalogue reference given together with shape and size of unit. Typically supplied with ironmongery. In this instance glazing units supplied separately. Note : bedding and pointing frames deemed included.
14	Glass, sealed double glazing units Comprising 4mm thick low emissivity glass (2nr) and 12mm gas filled airspace to timber frames and casements with non-setting mastic and wood beads, pane size 500 x 250mm, glazing rebates 20 -30mm	NRM2 23.8.1/2.2/3.3
2	Ditto; 460 x 250mm; glazing Rebates 20-30mm	NRM2 23.8.1/2.2/3.3 Use of the word 'ditto' here avoids the need to repeat the main body of the description immediately above. It means 'as has been said before' and is used here in combination with the only variable which, in this case, is the size of the glazing unit
2/ 1.77 1.20	Painting to glazed surfaces, base coat (1nr) Sikkens Cetol HLS > 300mm girth application on site prior to fixing, external	NRM2 29.2.2.6

Windows (Contd)

NRM2 29.2.2 and NRM2 29.2.1

1.77 1.20	Painting to glazed surfaces, base coat (2nr) Sikkens Cetol and finish coat Sikkens Filter 7 > 300mm girth, external & Ditto but internal

All decoration work requires a reference in the description as to whether it is internal or external. In this case the window requires decorating to both internal and external faces. An area is recorded for both instances by booking the window elevation area once (external) and then using an ampersand and 'ditto' to pick-up the decoration for the identical area but this time internally.

Opening Adjustment

1.77 1.20	*Deduct* Walls; 103mm thick brickwork in skins of hollow walls all as before described. & *Deduct* Forming cavities 80 mm wide inc. 4 nr. stainless steel wall ties/m² & *Deduct* Rigid Board cavity wall insulation 50mm thick abd & *Deduct* Finish to walls 13mm thick 2 coat plaster work to blockwork base > 600mm wide & *Deduct* Painting to general surfaces, basecoat (1nr) and finishing coats (2 nr) in emulsion paint to plastered surfaces >300mm girth, internal.

At this point an adjustment (deduction) is made for the previous over-measurement of the external walling and finishing trades. This is commonly referred to as the opening adjustment.

All of these items have previously been measured so it would be possible to include the use of the term 'as before described' (abd) in the description rather than repeat the full text.

The items that follow (next page) are also part of the opening adjustment. These are necessary in forming a window or external door opening but are all additions.

Windows (Contd)

Opening Adjustment (contd)

<u>Lintel Length</u>

Opening width	1 770
End bearing 2/150	300
	2 070

Nearest manufactured increment
<u>over =</u>   <u>2 100 mm</u>

1	Isolated metal member, galvanised IG steel insulated lintel type L1/S70; 2 100mm long, bedded in g.m. (1:4)	NRM2 14.25.1.1.1
1.77	Bands; 215mm x 65mm brick-on-end flush, horizontal, entirely of stretchers laid vertically in g.m. (1:4) with a flush joint	NRM2 14.7.1.3.1

Cross sectional area check
    $1.770 \times 0.225 = 0.41 \, m^2$

Deduction for displaced brickwork only made where the cross-sectional area > 0.50m². In this case no adjustment is required

Cavity tray   <u>length</u>

Lintel length	2 100
O'lap at ends 2/ 150	300
	2 400

	<u>width</u>
block height	225
allow for slope	50
build in ends 2/75	150
	425

2.40	Damp proof course width > 300mm horizontal as cavity tray bedded in g.m. (1:4)	NRM2 14.16.17.1.3
0.43		

Windows (Contd)

2/ 1.20	Extra over walls for opening perimeters, closing 80mm wide cavity with 150mm insulation blockwork	NRM2 12.1.2.

&

Damp proof courses abd < 300mm wide, vertical   NRM2 12.16.*.1

2/ 1.20 1.77	Finish to walls 13mm thick 2 coat plaster work to blockwork base < 600mm wide	NRM2 28.7.2

(Reveals)

(Soffit)

&

Angle bead, galvanised mild steel to suit 13mm thick two coat plasterwork fixed with plaster adhesive to block   NRM2 28.28.1.1/2

&

Painting to general surfaces, basecoat (1nr) and finishing coats (2 nr) in emulsion paint to plastered surfaces >300mm girth, internal.   NRM2 29.1.2.1

→ $x \, 0.15 =$ m^2

Window Board
Length	1 770
Ends 2/ 75	150
	1 920

<u>Windows (Contd)</u>

1.92	Medium Density Fibreboard window board 25 x 225mm rebated and rounded on one edge, secret fixed to blockwork	NRM2 22.5.1/2	

&

<u>sill dpc width</u>

$$103$$
$$80$$
building in $$\underline{25}$$
$$\underline{\underline{208}}$$

Damp proof course width < 300mm horizontal beded in g.m. (1:4)   NRM2 14.16.*.3

$\longrightarrow$ _____ x 0.21 = _____ m²

<u>Prime only girth to window bd</u>

Board face width   2/225 450
Board thickness   2/ 25   <u>50</u>
                              <u>500</u>

Painting to general surfaces, prime only (1nr) and >300mm girth, internal.   NRM2 29.1.2.1

$\longrightarrow$ _____ x 0.50 = _____ m²

Area class for window bd decs

1.920 x 0.225 = 0.432 m²

Therefore = isolated surface

1	Painting to general surfaces, undercoat (1nr) and finishing coats (2nr) full gloss finish on primed isolated surfaces only (1nr) and >300mm girth, internal.	NRM2 29.1.3.1

250 *Windows, doors and standard joinery*

2/ 1.20 1.20	Painting to glazed surfaces, base coat (1nr) Sikkens Cetol HLS > 300mm girth application on site prior to fixing, external	Before being built-in to the cavity wall the window unit receives a base coat of stain finish.
1.77 1.20	Painting to glazed surfaces, base coat (2nr) Sikkens Cetol and finish coat Sikkens Filter 7 > 300mm girth, external & Ditto but internal	Sometime later when the window has been fixed in place and forms part of the external envelope of the building, it receives a further three coats of stain finish. The external and internal work requires measuring separately, hence the use of the ampersand so that the previously booked internal area can be reused.

11.11 Internal door drawing and specification

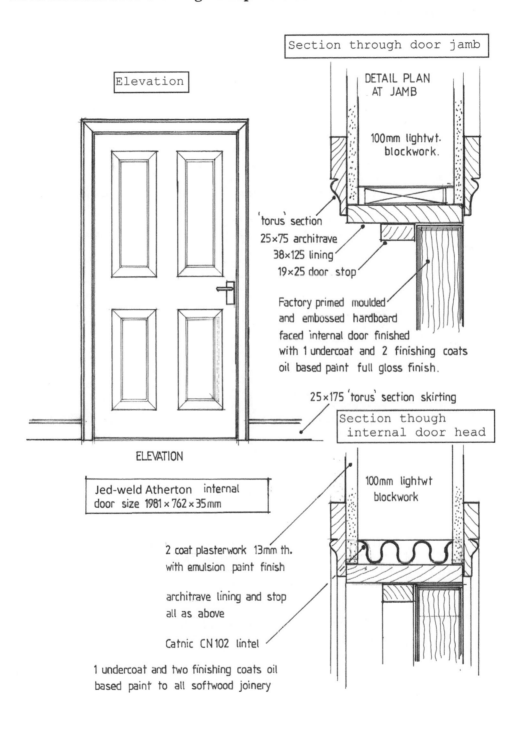

Elevation

Section through door jamb

DETAIL PLAN AT JAMB

100mm lightwt. blockwork.

'torus' section
25×75 architrave
38×125 lining
19×25 door stop

Factory primed moulded and embossed hardboard faced internal door finished with 1 undercoat and 2 finishing coats oil based paint full gloss finish.

25×175 'torus' section skirting

ELEVATION

Jed-weld Atherton internal door size 1981 × 762 × 35 mm

Section though internal door head

100mm lightwt blockwork

2 coat plasterwork 13mm th. with emulsion paint finish

architrave lining and stop all as above

Catnic CN102 lintel

1 undercoat and two finishing coats oil based paint to all softwood joinery

11.12 Internal door measurement example
(Job ref: SJIO 9597/ID)

Standard Internal door   Job ref: SJIO 9597/ID   Page 01 of 06

The work comprises the provision and installation of a standard internal door for a residential property all in accordance with floor plans, elevations and schedules provided

Take off list

STANDARD INTERNAL DOOR
NRM2 requires the following drawings to be available:-
General arrangement plans, Internal Door, Glazing and Ironmongery Schedules.

Mandatory information to be provided for windows :-
See NRM2 work section 24 for mandatory information to be provided .

Take-off list:-

- Internal door
- Door lining
- Architrave
- Door stop
- Decoration to door
  - To door + door edges
  - To door surround
- Ironmongery
- Opening Adjustment
  - Lintel
  - Blockwork
  - Plasterwork
  - Emulsion paint to wall
  - Skirting
  - Decs to skirting
  - Floor screed
  - Floor finish

1	Doors, 35mm thick factory primed hardboard faced moulded and embossed internal flush door, Jeld-Wen 'Atherton' 762 x 1981mm screw fixed with pair of door butts (measured separately) flush fixed to softwood door lining.

NRM2 24.2.1.*.1

An internal door schedule will help to identify the correct door specification and speed up the process of measurement . Care should be taken to check each door width and thickness for accessibility purposes, fire rating and glazing.

In some instances internal doors are supplied with ironmongery already in place and/or factory glazed. These should be enumerated and the description amended to reflect this.

Internal door (Contd)

Door lining dimensions
2/ 762   1 524
                          1 981
                          3 505
Add overlap at corners
              2/38   76
                        3 581

| | 3.58 |

Door lining... where doors are supplied complete with door linings they are measured as a 'door set' and enumerated as a composite item all in accordance with NRM2 24.1

Door lining 125 x 38mm wrot softwood levelled, plugged and secret screw fixed to blockwork

NRM2 24.10.1.1

Architrave length

Height   1 981
Add clearance   4
        mitre cut   75   79
                            2 060

Width   762
Add clearance 2/4   8
        mitre cut 2/75 150
                          920

2/2/ 2.06
2/ 0.92

Architraves 25 x 75mm wrot softwood 'torus' section including 4nr mitre cuts at angles

NRM2 22.2.1/2

Door stop
 Height of door   1 981
Less depth of stop   19
                          1 962

2/ 1.96
0.76

Door stop 19 x 25mm wrot softwood

NRM2 22.3.1

Decoration to door blank

Width of door   762
Decs to door edge   35
                         797

Internal door (Contd)

Decoration to door blank (Contd)

Height of door	1 981
Decs to door edge	35
	2 016

The area of the door blank is based on the width and height of the door face plus one door edge. The area is then times by two which will include both faces of the door plus the opening edges.

2/ 0.80
   2.02

Painting to general pre-primed surfaces, base coat (2nr) undercoats, finishing coat (1nr) gloss paint > 300 mm girth, internal

NRM2 29.1.2.1

Decoration to door surround joinery (see section sketch)

Length of decs to door lining

	(w)	762
2/38		76
		838
	(L)	1 981
		38
		2 019

Girth of door lining

2/125	250
2/ 38	76
	326

This is a waste calculation to shows the perimeter length and girth (width) of the door lining so that it can be prepared for decoration and primed before being fixed in position. In this case it establishes that it should be measured in m^2 (i.e. the girth exceeds 300mm)

Kps = knot, prime and stop

2/ 2.02
   0.33
   0.84
   0.33

Painting to general wrot softwood surfaces, knot prime and stop > 300 mm girth, internal, application on site prior to fixing

NRM2 29.1.2.1.6

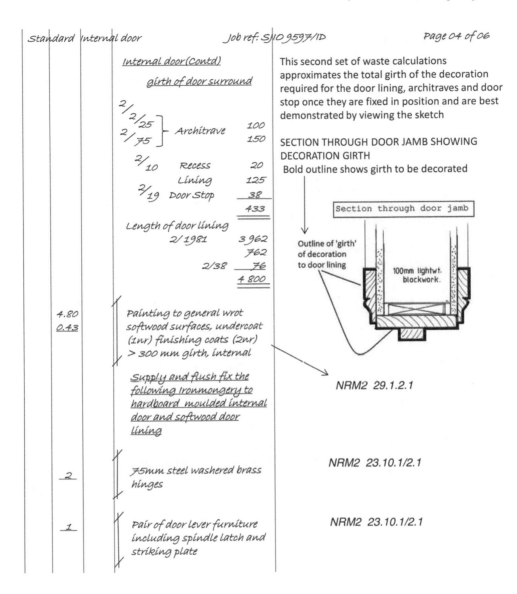

Internal door (Contd)

  girth of door surround

2/
2/
  /25 ⎤   100
2/   ⎦ Architrave
  /75   150

2/
  /10   Recess   20
         Lining   125
2/
  /19   Door Stop   38
                   ___
                   433
                   ═══

Length of door lining
    2/ 1981   3 962
                   762
    2/38   76
                 ─────
                 4 800
                 ═════

4.80   Painting to general wrot
0.43   softwood surfaces, undercoat
        (1nr) finishing coats (2nr)
        > 300 mm girth, internal

        Supply and flush fix the
        following Ironmongery to
        hardboard moulded internal
        door and softwood door
        lining

  2   75mm steel washered brass
        hinges

  1   Pair of door lever furniture
        including spindle latch and
        striking plate

This second set of waste calculations approximates the total girth of the decoration required for the door lining, architraves and door stop once they are fixed in position and are best demonstrated by viewing the sketch

SECTION THROUGH DOOR JAMB SHOWING DECORATION GIRTH
 Bold outline shows girth to be decorated

Section through door jamb

Outline of 'girth'
of decoration
to door lining

100mm lightwt.
blockwork.

NRM2 29.1.2.1

NRM2 23.10.1/2.1

NRM2 23.10.1/2.1

Internal Door (Contd)

Lintel length

door opening width		762
lining	2/38	76
end bearing	2/150	300
		1138

Nearest manufactured length over = 1 200mm

1

Isolated metal member galvanised IG steel lintel type INT100 1 200mm long, bedded in gauged mortar (1:4)

Lintels are manufactured in standard incremental lengths of 150 or 300 mm. Having established the minimum length required by adding end bearing to the opening width a lintel should be selected from the next available manufactured length **over** this minimum.

NRM2 14.25.1.1.1

Opening Adjustment

Height of door	1 981
add lining th	38
	2 019

width of door	762
add lining th 2/38	76
	838

0.84
2.02

Deduct
Walls, 100mm thick lightweight blockwork all as before described

An adjustment is necessary for the area of walling + plaster + decorations previously measured. In this instance the walling area is plastered and decorated on both faces so the blockwork area is measured and this base area timesed by two to adjust for the plaster and decoration on both faces of this block wall.

NRM2 14.1.2

&

Deduct
Finish to walls 13mm thick 2 coat plasterwork to block base > 600mm wide

NRM2 28.7.2

x 2 =   m²

&

Deduct
Painting to general surfaces basecoat (1nr) and finishing coats (2nr) in emulsion paint to plastered surfaces > 300mm girth internal

NRM2 29.1.2.1

x 2 =   m²

Standard Internal door   Job ref: SJO 9597/ID   Page 05 of 6

Internal Door (Contd)

Skirting adjustment
Door opening width   838
Architrave 2/75   150
                          988

2/ 0.99

Deduct
Skirtings 25 x 75mm wrot sw   NRM2 22.1.1
'torus' section all abd
        &

Deduct
Painting to general surfaces
basecoat (1nr) and finishing
coats (2nr) in emulsion paint   NRM2 29.1.2.1
to plastered surfaces >
300mm girth internal

→   x 0.08 =   m²

        &

Deduct
Painting to general wrot   NRM2 29.1.1.1.6
softwood surfaces, knot prime
and stop < 300 mm girth,
internal , application on site
prior to fixing   NRM2 29.1.1.1.

        &

Deduct
Painting to general wrot
softwood surfaces, undercoat
(1nr) finishing coats (2nr) <
300 mm girth, internal

This may seem counter intuitive but this is an adjustment for emulsion paint which has previously been deducted twice.   Once when the door opening adjustment was made and once again when the skirting was measured.

2/ 0.99
   0.18

Painting to general surfaces
basecoat (1nr) and finishing   NRM2 29.1.2.1
coats (2nr) in emulsion paint
to plastered surfaces >
300mm girth internal

Standard	Internal door	Job ref: SNO 9597/1D	Page 06 of 06

Internal Door (Contd)

Here an adjustment needs to be made for the additional floor area that has been created in forming the internal door opening. Previously floor finishes were measured based on the internal plan area of each room. Both the floor screed and the floor finish have previously been measured so the descriptions need only provide sufficient information to identify these (as before described).

Adjustment for floor screed and floor finish into door opening

Door threshold opening width
Blockwork 100
plaster 2/13 26
126

0.76
0.13

61mm thick cement and sand (1:4) screeds beds and the like > 600mm wide all as before described

NRM2 28.1.2.1.

&

4mm thick vinyl sheet finish to floors all as before described

NRM2 28.2.2

12 Floor, wall and ceiling finishes, dry linings, internal partitions and suspended ceilings

12.01 Introduction

The term Surface Finishes (NRM2 Work Section 28) is used to identify a number of finishing trades associated with the completion of the floors, walls and ceilings of a building. Each of these embraces several different operations including plastering, floor screeds, wall and floor tiling, sheet floor finishing, decorative papers or fabrics and decorating. Associated trades that are given their own NRM2 Work Section are Decoration (NRM2 Work Section 29) Suspended Ceilings (NRM2 Work Section 30) and Proprietary linings and partitions (NRM2 Work Section 20). Consideration of all of these trades will be included in this chapter.

12.02 Sequence of measurement

As with many other aspects of measurement, the key to an efficient and coherent take-off lies in a sensible subdivision of the work and a logical and consistent approach. Since individual circumstances and customs will vary, it is difficult to prescribe a specific approach to measurement. What follows therefore assumes a pattern of finishes that repeat themselves throughout a building. In practice, it is unlikely that a building would be finished in any other way. In any event, the preparation of a schedule will assist in identifying like with like, thereby avoiding a piecemeal, or room-by-room, approach (Figure 12.1).

When recording dimensions for finishing trades, there is the potential to measure some of the following items together (see [1] and [2] below). Adopting the following sequence of measurement is likely to save time when recording dimensions, by avoiding the need to repeat (the same) sets of dimensions twice.

FINISHINGS SCHEDULE		Project ref:			Job ref:				
	ceilings		floors		walls		skirting		notes
location	Base and finish	decs	Base and thickness (mm)	Floor finish	base	finish	type	decs	
reception	C1	D2	Sc 65	F3	P2	D4	Tb 1	SD4	
office 1	C1	D2	Sc 80	F2	P2	D2	Tb2	SD2	
office 2	C1	D2	Sc 80	F2	P2	D2	Tb2	SD2	
board rm	C1	D2	Sc 65	F3	P2	D4	Tb 1	SD4	
ladies wc	C2	D3	Sc55	F4	P3	D3	QT	-	
gents wc	C2	D3	Sc55	F4	P3	D3	QT	-	
workshop	C0	-	Sc G 85	FD1	P1	D1	-	-	

Ceilings	C0	No finish
	C1	12.5mm plasterboard with 3mm scim coat finish
	C2	12.5mm duplex plasterboard with 3mm scim coat finish
Decoration	D2	Mist coat + Two coat emulsion
	D3	Mist coat + Two coat eggshell emulsion
Floor screed	Sc	Screed (followed by thickness in mm)
Floors fin	FD1	Granolithic screed with power float finish
	F2	Medium duty vinyl sheet
	F3	Medium duty carpet on underlay
	F4	150 x 150 x 19mm quarry tile bedded, jointed and pointed in c&s (1:4)
Walls	P1	Fair face blockwork
	P2	2 coat plasterwork 13mm thick
	P3	2 coat c&s render 13mm thick (1:4)
Wall decs	D1	Sealing coat and two coats masonry paint
	D2	Mist coat + Two coat emulsion
	D3	150 x 150 x 4mm plain white glazed wall tiles (floor to ceiling) fixed with tile adhesive, finished with white grout
	D4	Wallpaper (PC sum £10/sqm)
Skirting	Tb1	150 x 25 mm 'torus' selected hardwood for varnish finish
	Tb2	150 x 25 mm bullnose softwood
	QT	150 x 19 x 100mm high coved quarry tile skirting with rounded top edge
Skirting decs	SD2	KPS, 2 undercoats, 1 finishing coat oil based paint, gloss finish
	SD4	Prepare and apply 3 coats matt finish varnish

Figure 12.1 Example finishing schedule.

- floor finishes and screeds;[1]
- ceiling finishes and decoration;[1]
- skirtings, coving, dadoes and decoration;[2]
- wall finishes and decoration.[2]

Working on the assumption that floor areas and ceiling areas are the same, the set of dimensions used to book floor finishes[1] can be linked by way of an ampersand to record ceiling finishes and associated decorations.[1] Similarly, where the internal perimeter of each room is established for the measurement of skirtings,[2] this same base dimension can be multiplied by the constant floor-to-ceiling height and used to provide the wall plaster area[2] and the wall decoration area.[2] Where possible, descriptions should be grouped in this fashion around a common set of dimensions. It is normal measurement practice to ignore window and door openings, recesses and other features when measuring finishes. In due course, an adjustment for this initial over-measurement will be made (see Chapter 11, Window and door opening adjustment).

12.03 General rules of measurement

As with other NRM2 Work Sections, the reader would be advised to refer to the Mandatory Information that must be provided, and also to note the works and materials that are deemed included (see NRM2 Rules for Detailed Measurement of Building Works 2.14, and the opening section of NRM2 Work Section 28). The following is an overview of these. Since all work is deemed internal, the measurer must state in a heading, or as part of the description, whenever work in this section is external. Work to attached beams is included with ceiling finishes, while work to attached columns is included with wall finishes. Finishes to isolated beams and columns must be given separately. Where the floor-to-ceiling height exceeds 3.50 m (as caused by atria, stairwells and the like), work to ceilings and beams has to be measured and described as over 3.50 m above the structural floor level. The principle unit of measurement for wall, floor and ceiling finishes is square metres. Where the width does not exceed 600 mm, the unit of measurement is linear. Individual widths are established on each face or surface. Where this is caused by attached beams and columns, it is classified as work to the abutting wall or ceiling and deemed included. No deduction is made for voids of less than 1.00 square metres, and all work associated with forming voids of less than 1.00 square metres is deemed included.

12.04 Wall, floor and ceiling finishes

12.04.01 Ceiling finishings

The ceiling area is measured in square metres between wall surfaces, stating in the description the type of finish and overall thickness. Where plasterboard or other sheet baseboard is specified, this should also be included in a single description followed by the thickness and number of skim coats. In cases where the ceiling finishes are identical throughout a floor or storey, it may be appropriate to measure gross (i.e. ignoring any partitions and internal walls) and at a later point to deduct the plan area of internal walls and partitions from the gross ceiling area (internal wall dimensions may well be available from earlier measurements). Care should be taken to identify and measure separately work in staircase areas, atria and plant rooms where floor-to-ceiling heights are likely to vary from the norm. Descriptions should include the details required by NRM2 28.9/10.1/2.1. Where a consistent sequence of measurement is adopted, all the dimensions relating to each ceiling finish description will be entered as a string of dimensions in the take-off. The need to establish a descriptive heading is therefore eliminated. Careful and consistent signposting is vital in order to avoid confusion at a later stage.

12.04.02 Floor finishings

Cement and sand screeds, together with floor finishes, are measured separately in square metres but only where their width exceeds 600 mm (NRM2 28.1.2). Work less than 600 mm wide should be measured in linear metres in accordance with NRM2 28.1.1. Work to floors must make reference in the description to the slope or plane of the finished screed (NRM2 28.1.1/2.1,2 or 3).

Sheet flooring, ceramic tiles, carpet or other floor tiles and edged fixed carpeting are likewise all measured in square metres providing they exceed 600 mm in width. In some situations it is possible to utilise the previously booked ceiling areas for the measurement of floors. The use of an ampersand to link these two sets of dimensions can save a great deal of time, but care should be taken to ensure the respective plan areas are consistent (Figure 12.2).

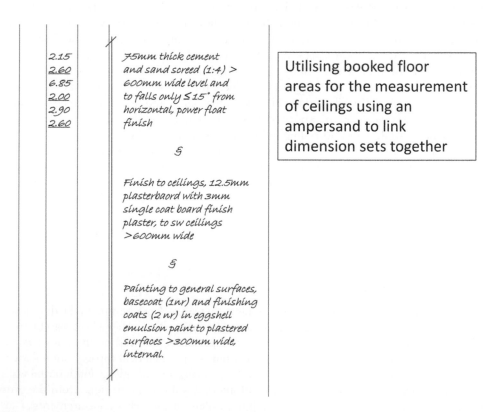

Figure 12.2 Typical finishing measured items as 'booked'.

Floor finishes through door openings will be measured as part of the door opening adjustment, as will dividing strips between different types of floor finishing (NRM2 28.35.1.1.1) (see Window and door opening adjustments, Chapter 11). Where floor finishes of different thicknesses abut at door openings, it is necessary to adjust the screed thickness in order to achieve a level floor finish throughout (Figure 12.3).

12.04.03 Wall finishings

This includes the wet trades of plastering and rendering to walls. For dry lining and similar work, please see Proprietary systems and dry linings (12.08). For plasterwork to walls, it is the area in contact with the base that is measured. This is recorded in square metres (to each wall face) where the width of the walling exceeds 600 mm; where the width of the plasterwork does not exceed 600 mm, it is measured in linear metres. The most efficient approach is to record the perimeter length of each room (established as a waste calculation) and then transfer the resultant perimeter length to the dimension column, where it is followed by the floor-to-ceiling height. Alternatively, individual room lengths can be recorded in linear metres and this string of linear measurement used in combination with an ampersand to group together the measurement of items that share the same common perimeter length of the room, such as skirtings, dado rails, picture rails (together with decorations to each), coving, wall finishes and wall decoration. The last two items will require multiplying by the room height (constant dimension) in order to convert them to an area. Signposts must be used to locate the base dimensions to their origin (Figure 12.4).

All ends and angles, together with rough and fair square cutting, are deemed included. Curved work is classified in the same fashion as work to flat walls (square metres if >600 mm wide and linear metres if ≥600 mm), but must state the radius of the wall in the description. Metal angle beads are measured in linear metres, giving a dimensioned description in accordance with NRM2 28.28 and stating the method of fixing and nature of the background in the description.

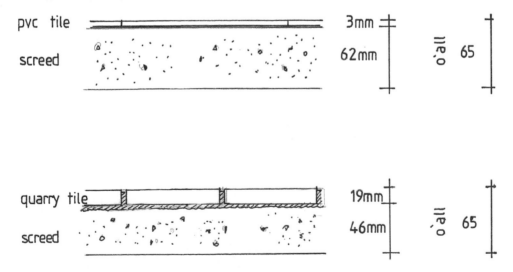

Figure 12.3 Adjusting screed depth for floor finish thickness.

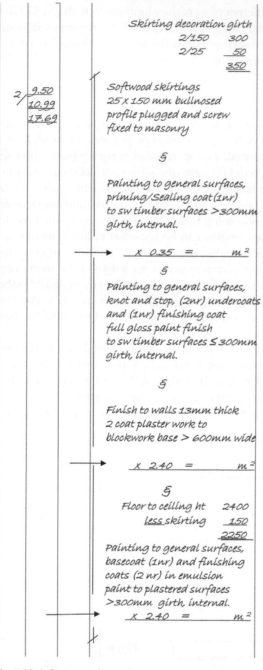

Skirting decoration girth
2/150   300
2/25   50
         350

2/ 9.50
   10.99
   17.69

Softwood skirtings
25 x 150 mm bullnosed
profile plugged and screw
fixed to masonry

&

Painting to general surfaces,
priming/sealing coat (1nr)
to sw timber surfaces >300mm
girth, internal.

x 0.35 =   m²

&

Painting to general surfaces,
knot and stop, (2nr) undercoats
and (1nr) finishing coat
full gloss paint finish
to sw timber surfaces ≤300mm
girth, internal.

&

Finish to walls 13mm thick
2 coat plaster work to
blockwork base > 600mm wide

x 2.40 =   m²

&

Floor to ceiling ht   2400
less skirting   150
                       2250

Painting to general surfaces,
basecoat (1nr) and finishing
coats (2 nr) in emulsion
paint to plastered surfaces
>300mm girth, internal.
x 2.40 =   m²

Individual room lengths (girths) can be recorded in linear metres and this string of linear measurement used in combination with an ampersand to group together the measurement of items that share the same common perimeter length (girth) of the room.

In this example there are three cases where this technique has been utilised:-

1. *Decoration to skirting*
2. *Wall plaster*
3. *Emulsion paint to wall*

Figure 12.4 Constant dimension.

Coving, moulding, bands, cornices and the like are measured in linear metres, giving a dimensioned description and stating the method of fixing and nature of the background in the description (NRM2 28.17–22). All work in forming ends, intersections, internal and external angles is deemed included (NRM2 28. 1–15, Mandatory Information).

12.05 Timber skirtings

Timber skirtings, picture rails and the like are measured in linear metres, and would usually be recorded for measurement purposes at the same time as plasterwork to walls or other wall finishings. All are based on room girths and can be 'anded-on' using an ampersand to the same set of dimensions that have been established for recording the wall plaster measurement (assuming the constant dimension approach is adopted – see Figure 12.4). The labour items of forming ends, angles, mitres and intersections are deemed included. It would be usual to include the measurement for the decoration of these timber components at this stage, and an explanation of how these are recorded is included later in this chapter (see Painting and decorating, below).

12.06 Painting and decorating

Painting and decorating merits its own separate Work Section (NRM2 29). The unit of measurement for painting and decorating is based on the girth of the surface receiving decoration. Where this exceeds 300 mm it is measured as an area; where it is an isolated surface and does not exceed 300 mm in girth, it is measured in linear metres; and when decorating isolated areas of less than 1.00 square metres, these should be enumerated, regardless of the girth (NRM2 29.1–7.1, 2 or 3). Paintwork to walls, ceilings, beams and columns is classified as work to general surfaces, and will appear in the BQ as a single conglomerate area (NRM2 29.1). Paintwork to glazed surfaces, structural metalwork, radiators, gutters, pipes and services should all be classified separately in accordance with NRM2 29.1–7. The unit of measurement will depend on the girth of the item and whether it is classed as an isolated area (see above). Paintwork to railings, fences and gates is treated in the same fashion, including reference as to whether the railing, fence or gate in question is closed, open or ornamental (NRM2 29.8). When adopting the group method of measurement, it is likely that decoration items will be included at the time of measuring the item being painted. For example, decoration items would normally be 'anded-on' with an ampersand to the measurement of eaves boarding when measuring roof finishes, to glazed doors when measuring internal doors and to radiators when measuring service installations. As a result of this approach, there is not normally a specific section in the take-off for paintwork since it is included with the item that is being decorated. As a result, the full extent of the decoration work will only become apparent once the completed BQ is established.

The 'girth of decoration' is an expression used to identify the sectional perimeter length of the surface being painted. For example, a timber component of sectional size 175 × 25 mm decorated on all faces (175 + 25 + 175 + 25) would give a girth of 400 mm. The same component fixed to a wall may only require decorating on the exposed face (175 + 25) and would give a painted girth of 200 mm. The distinction may seem arbitrary, but it is important since it determines the unit of measurement. Paintwork exceeding 300 mm girth is given in square metres, while work of less than or equal to 300 mm is described as 'isolated surfaces' and measured in linear metres. Because of this classification based on girth, it will be necessary to carry out a waste calculation before measurement can commence. A similar waste calculation may be necessary in order to identify isolated areas of less than 1.00 square metres. For example, a loft hatch size 600 × 900 mm (0.54 m²) would be measured as an enumerated item (general surfaces), while a panel radiator size 1800 × 600 mm (1.08 m²) would be measured in square metres and described as >300 mm girth and classified as 'painting radiators'. Where this distinction is obvious (for example, when measuring decoration to walls and ceilings), no waste calculation is necessary. It should be noted that it will be necessary to describe painting and decorating in NRM2 Work Section 29.1–7 as either internal or external (NRM2 29.1–7.*.1 or 2).

12.07 Decorative wallpapers/fabrics

The supply and hanging of decorative paper and fabric is included with decoration in NRM2 Work Section 29, and classified as work to walls and columns or work to ceilings and beams. Where areas of papering exceed 1.00 square metre these are measured in square metres, and where they are less than 1.00 square metres they are measured in metres (NRM2 29.9/10). Where appropriate, work on curved surfaces, lining paper and work to ceilings over 3.50 m high can be accommodated by providing a suffix to the description (NRM2 29./10.1/2.1; NRM2 29./10.1/2.*.1 and NRM2 29./10.1/2.*.2, respectively). No deduction is made for voids of less than or equal to 1.00 square metre. Border strips are measured in linear metres, whilst motifs are enumerated. All associated labours are deemed included.

12.08 Proprietary linings and partitions

12.08.01 Dry lining

Dry lining provides an alternative wall finish to the traditional two-coat plasterwork and is measured under NRM2 Work Section 20. Work to walls and ceilings should be described separately and measured on the exposed face in square metres, stating whether these surfaces are over 300 mm wide or not. In addition, the description should include any or all of the following as and where required: insulation, vapour barrier, sub-linings, finish, curved (stating radius), sloping and whether the lining is concave or convex (NRM2 20.10.*.1–7).

Dry lining to columns, beams and bulkheads is measured in linear metres, stating the girth in the description in one of the following four categories; not exceeding 300 mm, 300–600 mm, 600–900 mm and thereafter in 300 mm stages. In each of these categories it is also necessary to give the number of faces as part of the description (NRM2 20.12, 13, 14).

Forming openings for doors, windows and the like are measured as enumerated items, giving the size of the opening in the description in one of the following ranges: not exceeding 2.50 m², 2.50–5.00 m², exceeding 5.00 m² and thereafter in further increments of 2.50 m². Non-standard perimeter details such as deflection heads, acoustic seals and fire seals are measured in linear metres and described as 'extra-over' the work they are associated with, giving a dimensioned description or proprietary reference. The same approach should be adopted for any angles, junctions and finishing beads. The method of fixing and the nature of the base should also be included. Access panels are enumerated, giving a dimensioned description or proprietary reference.

12.08.02 *Metal-framed partitioning systems*

Increasingly nowadays, metal-framed partitioning systems are used as an alternative to timber stud partitions. For the purposes of measurement, these are grouped with dry lining systems (NRM2 Work Section 20.1–9). Systems to form internal walls are measured in square metres, stating the finished thickness and the height in one-metre increments along with the total length (based on the centre line) in the description. In addition, the description should include any or all of the following as and where required – insulation, vapour barrier, sub-linings, finish, glazing, curved (stating radius) – all in accordance with NRM2 20.1–3.*.1–7. No deductions are made for voids less than 1.00 m^2 (NRM2 20.1.*.*.*.1).

Metal-framed partitioning systems to ceilings are measured in square metres where the width exceeds 300 mm, and in linear metres where it is less than 300 mm. In similar fashion to the approach adopted for walling, the description should include any of the following as and where required: insulation, vapour barrier, sub-linings, finish, curved (stating radius), sloping, convex or concave, stating the radius (NRM2 20.2.1.1–7). The following items are all measured as 'extra-over' the work on which they occur (the unit of measurement is given in brackets following this): where the lining system finish changes (m^2), where there are openings (nr), non-standard perimeter details (m), angles (m), junctions (m) or access panels (nr). The exception to this is forming fair ends to partitions, which are measured in linear metres (m), giving the thickness of the partition in the description.

All work is deemed internal, and no deduction should be made for voids equal to or less than 1.00 m^2. Ends, fair ends and abutments with adjoining work, together with any additional framing to support heavy fittings such as radiators, wash handbasins and cisterns, are all deemed included (Figures 12.5 and 12.6).

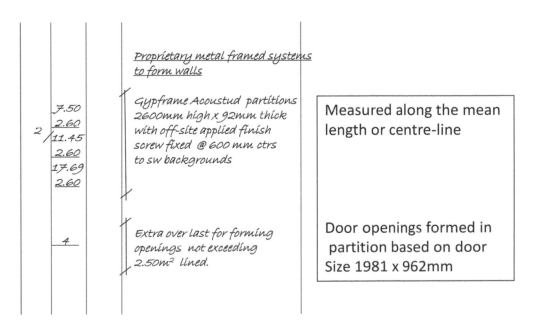

Figure 12.5 Proprietary metal stud partition.

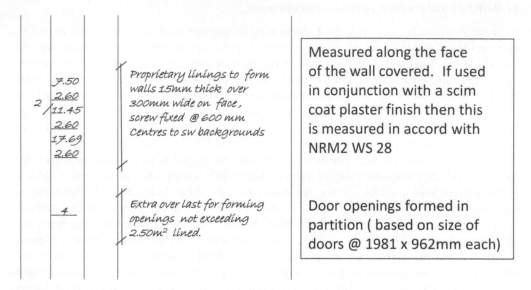

2	7.50		Proprietary linings to form
	2.60		walls 15mm thick over
	/11.45		300mm wide on face,
	2.60		screw fixed @ 600 mm
	17.69		centres to sw backgrounds
	2.60		

Measured along the face of the wall covered. If used in conjunction with a scim coat plaster finish then this is measured in accord with NRM2 WS 28

4 — Extra over last for forming openings not exceeding 2.50m² lined.

Door openings formed in partition (based on size of doors @ 1981 x 962mm each)

Figure 12.6 Dry lining.

12.9 Suspended ceilings

Suspended ceilings are measured in square metres on their exposed face, stating whether these are to ceilings, plenum ceilings, beams or bulkheads (Figure 12.7).

It is necessary to give the depth range of suspension in one of three categories: less than or equal to 150 mm, 150–500 mm and thereafter in stages of 500 mm (NRM2 30.1–4.1–3). In addition, the thickness of the tile and the method of fixing to the structure must be given together with any integral insulation and/or the inclusion of a vapour barrier. Where the height of the work exceeds 3.50 m above finished floor level, it should be given in 1.50 m stages (NRM2 30.*.*.1–5). Upstands, edge trims, angle trims, shadow gap battens and isolated strips are all measured in linear metres (NRM2 30.5, 6, 8, 9 and 13). Fire barriers can be measured in either square or linear metres (NRM2 30.10), while fittings can be measured as enumerated items or in linear metres (NRM2 30.12). Access panels (NRM2 30.7) and collars (NRM2 30.11) are enumerated (Figure 12.8).

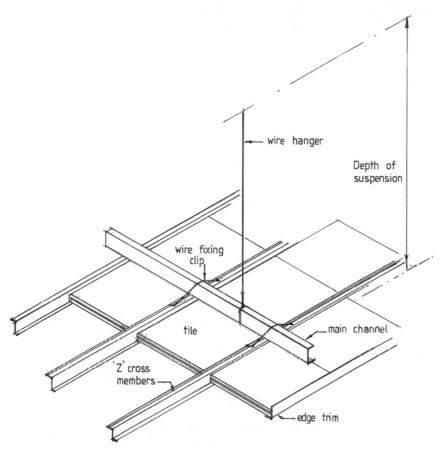

Figure 12.7 Suspended ceilings.

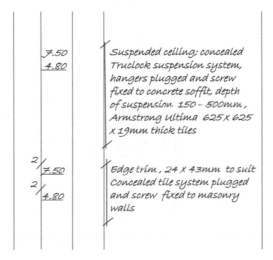

Figure 12.8 Suspended ceilings.

12.10 Internal finishes drawings, and specification (for a light industrial workshop and office)

Key for finishes:

Ceilings C0 No finish.
C1 12.5 mm plasterboard with 3 mm skim coat finish. Two coats emulsion.
C2 12.5 mm plasterboard with no finish.
C3 12.5 mm Duplex plasterboard with 3 mm skim coat finish. Two coats eggshell emulsion.
Floors F1 75 mm thick granolithic screed (1:4) with power float finish. Base sealing coat and two further coats Flowshield SL
F2 75 mm thick cement and sand screed (1:4) floated finish. Heavy duty carpet (edge fixed) on Duralay heavy duty underlay.
F3 75 mm thick cement and sand screed (1:4) floated finish. Medium duty carpet (edge fixed) on Duralay medium duty underlay.
F4 150 × 150 × 19 mm heather brown quarry tile, bedded, jointed and pointed in cement and sand (1:4).
F5 75 mm granolithic cement and sand screed (1:4) floated finish.
Walls W1 Sealing coat and two coats masonry paint to fair faced blockwork.
W2 Two coat plasterwork 13 mm thick, base coat and two further coats emulsion.
W3 13 mm thick cement and sand render (1:4). 150 × 150 × 4 mm plain white glazed wall tiles (floor-to-ceiling) fixed with adhesive and grouted with white cement.
Skirtings S0 No skirtings.
S1 150 × 25 mm softwood bullnosed. Knot prime and stop, two undercoats, one finishing coat full gloss finish oil-based paint.
S2 150 × 19 × 100 mm high coved quarry tile skirting with rounded top edge.

Room		Ceiling	Floor	Wall	Skirting
01	Workshop	C0	F1	W1	S0
02	Entrance foyer	C1	F2	W2	S1
03	Reception	C1	F3	W2	S1
04	Male WC	C3	F4	W3	S2
05	Female WC	C3	F4	W3	S2
06	Store	C2	F5	W1	S0
07	Manager	C1	F3	W2	S1
08	Corridor	C1	F2	W2	S1

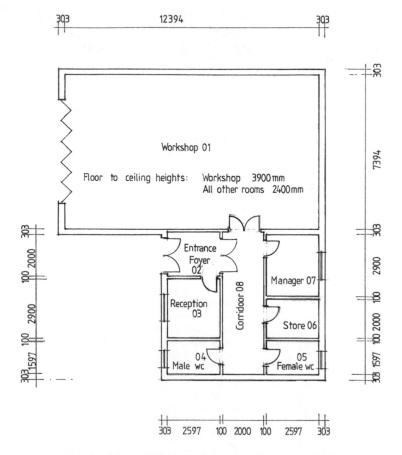

Concrete subfloors, blockwork walling, softwood joisted ceilings

12.11 Internal finishes measurement example
(Job ref. SOswldBay OL15 81/80)

Floor Wall & Ceiling Finishes   Job ref: SOswld Bay OL15 81/80   Page 01 of 08

NRM2 requires the following drawings to be available:-
General arrangement plans, Internal Door, Glazing and Ironmongery Schedules.

Mandatory information to be provided for floor, wall, ceiling and internal decoration :-
See NRM2 work sections 28 and 29 for mandatory information to be provided .

* Floor to ceiling heights all as detailed on drawings
* Assume concrete subfloors, blockwork walling and softwood joisted ceilings
* Note specification change F1 paint finish to granolithic screed now reads *Flowshield SL*
  in lieu of Sadolins floor paint

  Take off list

  * Floors
    * Screeds
    * Floor finishes
  * Ceilings
    * Plasterboard
    * Decoration
  * Walls
    * Wall finishes
    * Decoration
  * Skirtings
    * Timber skirtings
    * Decoration to ditto
    * Quarry tile skirtings

  Note:- workshop area, direct decoration to blockwork walling
          workshop area has no ceiling finish and no skirting

<u>Floors F1 – workshop (01)</u>

12.39	75mm thick granolithic cement and sand screed (1:4) > 600mm wide level and to falls only ≤ 15° from horizontal, power float finish (01
7.39	

NRM2 28.1

&

Painting to general surfaces, Flowshield SL 1 nr. Sealing coat and 2 nr. Finishing coats > 300 mm girth, internal

NRM2 29.1

<u>Corridor length (08)</u>

	2000
	2900
	1597
Partitions 2/100	200
Cavity wall (int)	303
	7000
Less 103 bwk	
50 cav	153
	6847

<u>Entrance Foyer (02)</u>

	2000
Cavity wall (int)	303
	2303
Less 103 bwk	
50 cav	153
	2150

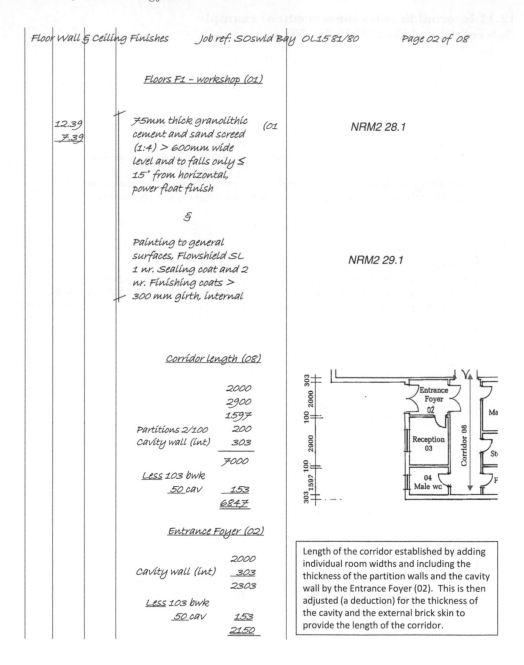

Length of the corridor established by adding individual room widths and including the thickness of the partition walls and the cavity wall by the Entrance Foyer (02). This is then adjusted (a deduction) for the thickness of the cavity and the external brick skin to provide the length of the corridor.

Floor Wall & Ceiling Finishes Job ref: SOswld Bay OL15 81/80 Page 03 of 08

			Floors – F2 (Entrance foyer)		
	2.15 2.60 6.85 2.00		75mm thick cement and sand screed (1:4) > 600mm wide level and to falls only ≤15° from horizontal, floated finish	(02 (08	NRM2 28.1
			&		
			Heavy duty carpet finish to screed floors including heavy duty underlay > 600mm wide, overall 24mm thick including perimeter fixing		NRM2 28.2
			Floors – F2 (Reception & Manager)		
2/	2.90 2.60		75mm thick cement and sand screed (1:4) > 600mm wide level and to falls only ≤15° from horizontal, floated finish	(07/03	NRM2 28.1
			&		
			Medium duty carpet finish to screed floors including heavy duty underlay > 600mm wide, overall 24mm thick including perimeter fixing		NRM2 28.2

274 *Floor, wall and ceiling finishes*

<u>Floors – F4</u>
    (04 and 05)

Screed thickness
O'all   75 mm
Quarry tile   <u>19 mm</u>
Screed depth <u>56 mm</u>

2/ 2.60 <u>1.60</u>	56 mm thick cement and sand screed (1:4) > 600mm wide level and to falls only ≤ 15° from horizontal	(04 (and (05	NRM2 28.1

                                                  &

	Heather brown quarry tile 150 x 150 x 19mm thick floor finish > 600mm wide pointed in cement mortar (1:4)		NRM2 28.2

Note:- too take; QT skirting

<u>Floors – F5</u>
    (Store)

2.60 <u>2.00</u>	75mm thick granolithic cement and sand screed (1:4) > 600mm wide level and to falls only ≤ 15° from horizontal, floated finish	(06	NRM2 28.1

<u>Ceilings – C1</u>
(Corridor, entrance hall, reception
and manager's office)

2.15	Finish to ceilings, 12.5mm (02	NRM2 28.9
2.60	plasterboard with 3mm	
6.85	Single coat board finish (08	
2/ 2.00	plaster, to sw ceilings	
2.90	>600mm wide (03/07	
2.60		

&

Painting to general surfaces,
basecoat (1nr) and finishing
coats (2 nr) in emulsion
paint to plastered surfaces
>300mm wide, internal.   NRM2 29.1

<u>Ceilings – C2</u>
(Store)

2.00   Finish to ceilings, 12.5mm   NRM2 28.9
2.60   plasterboard to sw ceilings (06
       >600mm wide

<u>Ceilings – C3</u>
(Male/Female WC)

2/ 1.60 Finish to ceilings, 12.5mm   NRM2 28.9
2.60   plasterbaord with 3mm (04/05
        single coat board finish
        plaster. to sw ceilings
        >600mm wide

&

Painting to general surfaces,   NRM2 29.1
basecoat (1nr) and finishing
coats (2 nr) in eggshell emulsion
paint to plastered surfaces
>300mm wide, internal.

Floor Wall & Ceiling Finishes   Job ref: SOswld Bay OL15 81/80   Page 06 of 08

Wall Finishes (W1)
Workshop (01)*
Store (06)

Room girth (01)
2/ 12 394   24 788
2/ 7 394   14 788
                39 576

Room girth (06)
2/ 2 000   4 000
2/ 2 597   5 194
                9 194

* Note NRM2 28.9.2.1
Workshop wall and ceiling height
exceeds 3.50m and therefore requires
a separate measurement and description

No finish necessary to workshop walls (F0).
Worth checking to see if any fair face finish
is required to the internal face of block
walling (would normally be measured with
Masonry Work).

39.58	Painting to general surfaces,	(01
3.90	sealing coat (1nr) and	
9.19	finishing coats (2nr)	(06
2.40	with masonry paint to	
	blockwork surfaces >300mm	
	wide, internal.	

NRM2 29.1

Wall Finishes (W2)
    and skirting (S1)

(02) Entrance Foyer
(03) Reception
(07) Manager
(08) Corridor

Room girths
(02)   2/ 2 597   5 197
         2/ 2 150   4 300
                        9 497
(03) 07)
         2/ 2 900   5 800
         2/ 2 597   5 194
                        10 994

(08)   2/ 6 847   13 694
         2/ 2 000   4 000
                        17 694

Skirting decoration girth

	2/150	300
	2/25	50
		350

2	9.50	Softwood skirtings 25 x 150 mm bullnosed profile plugged and screw fixed to masonry	NRM2 22.1
	10.99		
	17.69		

§

Painting to general surfaces, priming/Sealing coat (1nr) to sw timber surfaces >300mm girth, internal.   NRM2 29.1

x 0.35 =   m²

§

Painting to general surfaces, knot and stop, (2nr) undercoats and (1nr) finishing coal full gloss paint finish to sw timber surfaces ≤ 300mm girth, internal.   NRM2 29.1

§

Finish to walls 13mm thick 2 coat plaster work to blockwork base > 600mm wide   NRM 28.7

x 2.40 =   m²

§

Floor to ceiling ht	2400
less skirting	150
	2250

Painting to general surfaces, basecoat (1nr) and finishing coats (2 nr) in emulsion paint to plastered surfaces >300mm girth, internal.   NRM 2 29.1

x 2.25 =   m²

278 *Floor, wall and ceiling finishes*

<u>Wall Finishes (W3)</u>
<u>and skirting (S2)</u>

(04) Male WC
(05) Female WC

Room girths
(02)   2/ 2 597   5 197
        2/1 597   3 194
                     8 388

2/ <u>8.39</u>   Finish to walls 13mm thick   NRM2 28.7
        2 coat plaster work to   (04/05
        blockwork base > 600mm wide

        → x 2.40 =   m²

                  &

        Finish to walls 150 x 100 x 19mm   NRM2 28.14
        heather brown quarry tile skirting
        net height 100mm pointed in
        cement mortar (1:4)
             Floor to ceiling ht   2 400
             <u>less</u> skirting   100
                                   2 300

                  &

        Finish to walls 150 x 150 x 4mm
        white glazed wall tiles, plain, fixed   NRM2 28.7
        with adhesive to cement and sand
        render backing, grouted with white
        cement > 600 mm wide

        → x 2.30 =   m²

13 Drainage above ground, electrical and plumbing installation

13.01 Introduction

New domestic property will require a piped or cabled supply for the following services:

- lighting and power;
- hot and cold water;
- heating;

- waste disposal;
- communication;
- safety and security.

In non-domestic situations it is likely that a service engineer would be employed to provide detailed design and layout drawings, which would be used as the base to prepare quantities. In such cases the service installation work can be extensive and is likely to be given the title 'mechanical and electrical installation'. In regard to small-scale, low-rise residential property, the appointment of a mechanical/electrical service engineer is unlikely and limited design detail would be available. General layout drawings will identify the location of sanitary appliances, socket outlets, light switches, boilers, radiators and the like, but there will be no indication of pipe or cable runs; these will be left for the plumber or electrician to determine.

Due to the specialist nature of service installations, this work is often the subject of a Prime Cost Sum. While the measurement of domestic services is relatively straightforward, there is a presumption that the measurer has an understanding of the principles associated with services technology.

The measurement rules for the installation of services (NRM2) have been drafted to accommodate many different building functions (offices, hospitals, retail outlets). As such, the rules may appear overly complex when applied to the measurement of plumbing and electrical work for low-rise residential property. Another way to look at this would be to consider the measurement of plumbing and electrical services for a residential property as a 'safe first step' by way of an introduction to the measurement of mechanical and electrical installation in more complex buildings.

The rules for the measurement of plumbing and electrical services and the classification of the same work are found in the following NRM2 work sections:

32   Furniture, fittings, and equipment;
33   Drainage above ground;
38   Mechanical services;
39   Electrical services;
41   Builders' work in connection with mechanical and electrical installation.

In addition to the above, where the likes of lifts or hoists are required in a residential property, these should be measured in accordance with NRM2 Work Section 40 Transportation.

Where service installation drawings are provided, measurement in accordance with these rules is possible. Where no such detail exists, a Prime Cost Sum may be included on the assumption that this work will be carried out by a named subcontractor. The inclusion of Prime Cost Sums is given further consideration in Chapter 4.

Assuming sufficient drawn detail is available, dimensions can be booked. It is suggested that some form of logic is followed when recording dimensions. One approach is to follow the sequence of each service, starting with the entry from the site boundary and then following each service in and around the building. Rather than providing a full take-off, a take-off list and a set of descriptions based on commonly encountered installations has been provided (see 13.10 and 13.11).

13.02 Service entry to domestic buildings

The provision of water, gas, electricity and communication services is normally made in a single service trench, which terminates in a temporary access chamber near to the building. From this position, individual services are ducted to their different meter positions or entry points (Figure 13.1).

Service trenches are measured in linear metres in accordance with NRM2 41.13 (work for external service installation). The description should include the type and nominal size of the service, together with the average depth of the run to the nearest 500 mm. Where the service trench carries multiple services, this should be so described (NRM2 41.13.3). Although not specifically mentioned, it is assumed that earthwork support together with backfilling and disposals are deemed included. Utility companies will provide ducts for the contractor to install in the service trench, and at a later stage the services will be sleeved through these ducts and a connection made to the property. Separate enumerated measurement is required for pipe duct fittings together with any pipe accessories associated with the service entry (NRM2 41.15 and NRM2 41.16). The only exception to this is the water connection, which normally requires no sleeve (Figures 13.2 and 13.3).

The provision of each of the services to the point of metering is the responsibility of the appropriate utility provider. A standard charge will be made by each utility provider for a connection, and any payment will normally be required before work can commence. The cost of this is usually included in the Bill of Quantities by way of a Provisional Sum based on a quotation from the utility company. The gas, water, communication and electricity companies will normally provide meter cabinets or junction boxes together with sleeved rest bends for building into the cavity wall. An enumerated item should be included for forming and building in meter cabinets, giving details in the description of size, fixings, damp proofing, lintel support and any associated service ducting (NRM2 41.16). Where it is necessary for any service pipework to pass through the external wall, an allowance for lintel support to the substructure should be made (Figure 13.4).

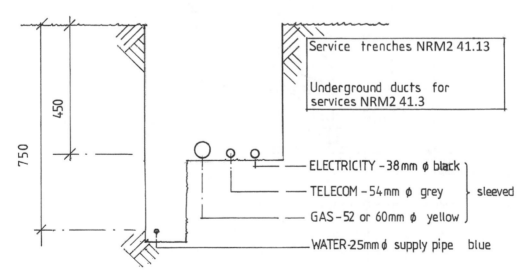

Figure 13.1 Section through service trench showing typical shared service entry for residential property.

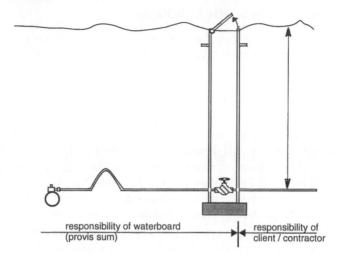

Figure 13.2 Mains connection (tapping) to water service.

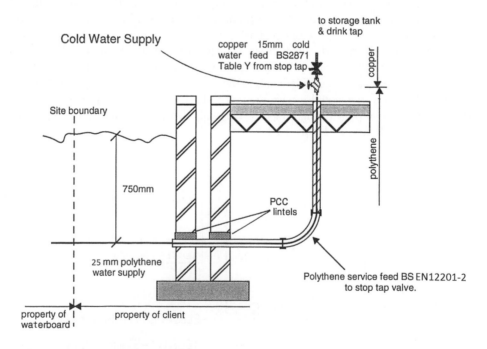

Figure 13.3 Mains water service entry to residential building.

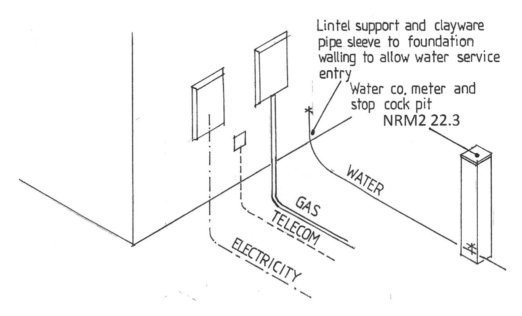

Lintel support and clayware pipe sleeve to foundation walling to allow water service entry

Water co. meter and stop cock pit NRM2 22.3

WATER

GAS

TELECOM

ELECTRICITY

Figure 13.4 Service entry to building (residential low rise).

In order to make service connections, it may also be necessary to include some provision for breaking up and reinstating the public highway. In the normal course of events this will be carried out by the appropriate service company and included as part of their standard charge. At a later stage the completed service installation will be tested and, subject to the public utility provider's approval, a connection to the service made (NRM2 41.26 and 41.27).

The following section identifies the rules of measurement for the services most commonly encountered in low-rise detached residential buildings.

13.03 Electrical installation

Before measuring the electrical installation associated with low-rise residential accommodation, it is worth while spending some time becoming familiar with common wiring practice and the graphical symbols adopted to identify electrical components and appliances. The following is offered by way of introduction to the measurement of the electrical installation required in residential property. For more complex installations, reference should be made to the regulations governing electrical installation published by the Institution of Electrical Engineers (IEE).

Electricity can only be transmitted through a conductor when there is a complete circuit from the source, via a conductor, back to the source. Each conductor cable contains a 'live' wire carrying the power to an appliance, a 'neutral' wire carrying the power back to the source and an 'earth' wire that reduces the risk of shock by carrying the current to a circuit breaker or to the ground in the event of a short circuit. The conductor used for domestic supplies (UK) is copper wire. In the UK, electricity is generated and supplied in three-phase AC (alternating current). For domestic installation only, a single-phase supply is required and this is delivered to the property via a service cable which terminates in a main fuse mounted in the meter cabinet. From here the connection to the meter is made and it is at this point that the responsibility of the electricity service provider ends. The consumer unit mounted inside the property steps down the supply ampage by dividing the incoming service into a number of separately fused circuits. Each circuit is protected by its own residual current circuit breaker (RCCB or MCCB), which detects any abnormal flow or surge of electricity and immediately cuts the supply (Figure 13.5).

There are normally three conductors in a cable – a live, a neutral and an earth. The live and neutral are insulated with brown and blue PVC, respectively, while the earth wire is uninsulated. All three wires are then sheathed in an outer layer of insulation. These are consequently known as PVC insulated twin and earth cables (T & EC). Cables are identified by the cross-sectional area of the conductors and this is expressed in square millimetres (mm^2). The larger the area of the conductors, the bigger the current that can safely flow through them. The number of wiring circuits (known as 'ways') found in a domestic installation will vary depending on the size of the property; typically there would be between ten and sixteen different circuits in the consumer unit for a residential property. The more common cable sizes for UK residential installation are shown in Table 13.1 and Figure 13.6.

As was noted earlier, NRM2 work section 39 provides the measurement classification for electrical installation. In order to reflect current regulation and best practice, it is recommended that reference is made to the most recent edition of the IEE Wiring Regulations. The following descriptions and approaches adopt the terminology used in the classification of measurement for electrical services in accordance with NRM2.

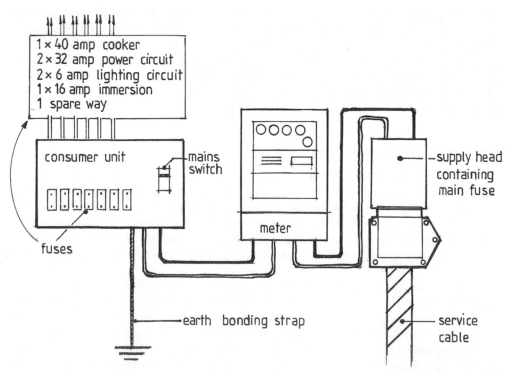

Figure 13.5 Mains electrical supply and consumer unit.

Table 13.1 Common electrical cable sizes for UK residential installation.

Cable size (mm²)	circuit	type	distribution
1.00mm²/1.50mm²	lighting	TC&E	per floor
2.50mm²	power	TC&E	per floor
2.50mm²	immersion heater	TC&E	single circuit
6.00mm² or 10.00mm²	cooker	TC&E	per appliance
6.00mm² or 10.00mm²	spontaneous shower	TC&E	single circuit
16.00mm²	earth cable	single core	earth bonding
16.00mm²	meter leads	single core	meter

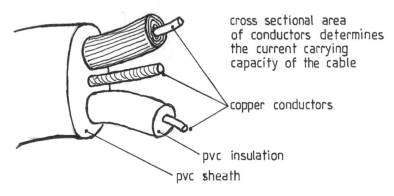

cross sectional area
of conductors determines
the current carrying
capacity of the cable

copper conductors

pvc insulation

pvc sheath

Figure 13.6 Twin and earth insulated cable.

13.03.01 Work in final circuits

Most electrical installation for single residential property will be included under this category. Ring mains, lighting circuits and cooker and immersion heater points should all be measured under this category. Each circuit is enumerated (nr), stating in the description the type and size of cabling and the number of socket outlets or lighting points serviced by the circuit, together with an indication of the installation location (e.g. roof space; NRM2 39.7.1.1). See also 13.11 Example take-off, page 2 of 10.

13.03.02 Power supply and socket outlets

These circuits are generally arranged as a ring main. A single 2.50 mm² twin and earth cable is looped in and out of each socket outlet and is then returned to the same fuse, thereby completing the circuit and forming a ring. Under normal circumstances a separate ring would serve each floor (Figure 13.7).

13.03.03 Lighting circuits

Lighting circuits are not installed as rings, although they are still measured as final circuits. Lighting circuits are normally installed with a 'loop-in' arrangement. Cabling runs directly from the fuseboard to a ceiling rose, which has four terminals. One of these connects to the ceiling rose, one to the switch, one to the lampholder via a flex and one to the next ceiling rose in the

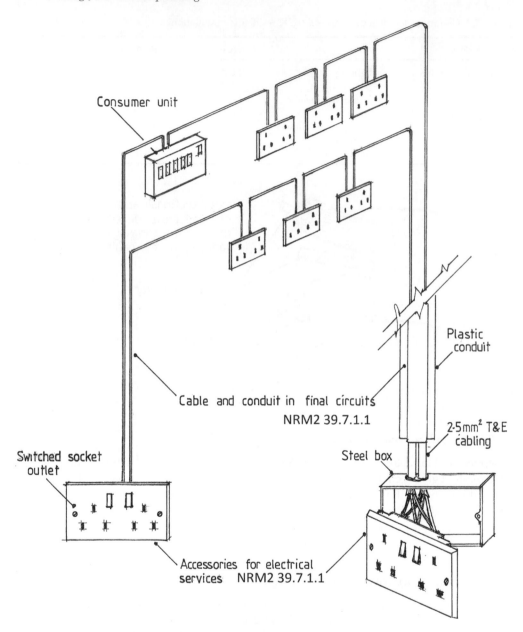

Figure 13.7 Schematic view of power supply (ring main).

circuit. Up to twelve light points can be installed on a single circuit in this way. It is standard wiring practice to fuse and wire separate circuits for each floor so that the failure of one does not affect the other.

13.03.04 Appliances

Separately fused circuits will be necessary for the cooker, immersion heater, electric heating units and spontaneous water heater or shower.

13.03.05 Work not in final circuits

Mostly found in non-domestic situations where a supply is brought into the premises and then fed to different locations in a building, where it is phased down through a fuse-box to supply individual appliances or final circuits (see above). Cable containment and cable are measured separately in linear metres (NRM2 39.3 and NRM2 39.5).

13.03.06 Concealed wiring and conduit installation

Where cables run behind plastered or dry-lined walling they will be run in conduit. In domestic situations this plastic sleeving prevents the cabling becoming permanently plastered into the wall and allows the cables to be replaced in the event of a fault. Conduit in the final circuit of a domestic installation is included as part of the description accompanying the enumeration of the cabling installation (see 13.11 Example take-off, page 2 of 10).

13.03.07 Builders' work in connection (BWIC)

The term 'builders' work in connection' is used to describe a number of operations including cutting holes, forming chases and sinkings, all of which are required to allow the completion of an electrical or plumbing installation. Historically this would have been carried out by the builder's labourer rather than the electrician or plumber – hence the term. Based on the presumption that minimum design information will be available, NRM2 includes a specific work section for the inclusion of an item to cover the measurement of BWIC (NRM2 41.1.1). However, it will be necessary to repeat this for each type of electrical installation (e.g. power circuit, lighting, immersion heater etc.). Before any cutting or chasing can begin, it will be necessary to mark the position of the proposed installation in and around the structure of the building. In similar fashion, NRM2 provides for measurement and costing by way of an item for this work (NRM2 41.2). As previously explained, this will also be necessary for each type of installation. These two inclusions provide sufficient information for measuring and costing purposes, and presume that the extent of the work is limited to smaller-scale installations, as would be the case for a single residential property. It should also be noted that work to new buildings and existing buildings needs to be measured and described separately. Finally, an inclusion must be made by way of an item for the testing and commissioning of the completed electrical installation (NRM2 39.15, NRM2 39.16).

A typical domestic electrical installation (take-off list) might include the items shown in Table 13.2.

Table 13.2 Items in a typical domestic electrical installation.

Item	Unit	NRM2 ref.
Underground service runs	m	41.13
Items extra over underground service runs	m,m^2,m^3	41.14
Building in flush electricity meter cabinet. Lintel over, dpc and polythene surround with preformed service tube	nr	14.25
Distribution board; primary equipment	nr	39.1.1
Cable and conduit in final circuits: switch sockets	nr	39.7.1.1
Cable and conduit in final circuits Immersion heaters etc.		
Cable and conduit in final circuits Lighting outlets		
General Builder's Work in Connection with lighting installation	Item	41.1.1
Marking positions of holes and chases etc. for lighting	Item	41.2
General Builder's Work in Connection with power installation	Item	41.1.1
Marking positions of holes and chases etc. for power circuits	Item	41.2
General Builder's Work in Connection with cooker installation	Item	41.1.1
Marking positions of holes and chases etc. for cooker installation	Item	41.2
General Builder's Work in Connection with immersion heater installation	Item	41.1.1
General Builder's Work in Connection with spontaneous shower installation	Item	41.1.1
Marking positions of holes and chases etc. for spontaneous shower	Item	41.2
Marking positions of holes and chases etc. for immersion heater	Item	41.2
Testing	Item	39.15
Commissioning	Item	39.16
Operation and maintenance manuals	Item	39.18

The following sample measured items relate to no specific example. The descriptions and quantities are typical of those normally encountered in a domestic electrical installation (see 13.11 Sample take-off, pages 1 to 10).

13.04 Plumbing and heating installation

In residential buildings this will involve a number of different systems associated with the supply, distribution and disposal of water in and about a building. There is no specific work section with the title 'Plumbing Installation' included in NRM2. Whenever measurement is necessary, it will need to draw on a number of related NRM2 work sections, including the following:

32 Furniture, fittings and equipment;
33 Drainage above ground;
38 Mechanical services;
41 Builders' work in connection.

As with any other work section, in order to measure the different components that together constitute a residential plumbing and heating installation, it will be necessary to have some understanding of plumbing and heating system service technology. Typically, for residential property, the design information available would not show intended pipe runs but would give the location of the principle equipment (boiler, hot water cylinder, sanitary appliances).

Providing the measurer is familiar with the installation system adopted, the process of recording dimensions is relatively straightforward. The design of the plumbing and heating installation will be influenced by a number of factors including: the availability or roof/loft space, the water pressure, the adoption of energy-saving technologies (solar panels and heat pumps) and the anticipated lifestyle of the end-user. Two alternatives are now described.

13.04.01 Stored water systems (indirect)

This approach involves the provision of a cold water storage tank (usually located in the roof space). Cold water is drawn directly from the water mains into the cold water storage tank and to the kitchen cold water tap. Cold water is then drawn indirectly (by gravity) to bathrooms, cloakrooms and en-suite bathrooms. A cold feed supply is also required for the hot water cylinder and boiler. A closed circuit of pipework between the boiler and the hot water cylinder ensures that there is a hot water supply available for bathrooms and the like (Figure 13.8).

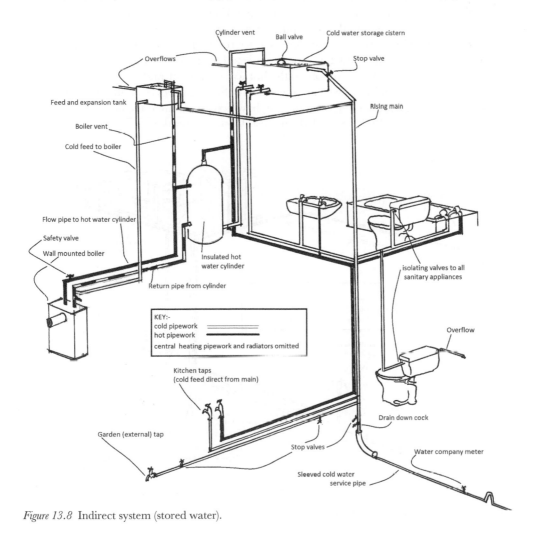

Figure 13.8 Indirect system (stored water).

13.04.02 Mains-fed water systems

Where mains water pressure is sufficient, a mains-fed system can be installed. In this system the cold water storage tank is eliminated and cold water is supplied to all appliances directly from the mains. Hot water is delivered via a combination boiler, which only heats water as and when it is required. Mains-fed systems are typically unvented and operate under pressure (Figure 13.9).

Once the appropriate system has been identified, it is suggested that the measurer adopts a logical approach to booking dimensions by following the flow of water through the property. In the first instance this would mean bringing both water and gas services from the appropriate mains (usually in the road) to the property.

13.04.03 Cold water installation

Household water supply enters the property via the water company's service pipe (as described earlier in this chapter) and terminates at a stop valve. From this point it rises to a cold water storage cistern usually located in the roof space. A ball valve in the cistern controls the supply of water flowing under pressure from the local service reservoir. In addition to the rising main inlet, there are generally three further pipes connected to the cold water cistern. These are the overflow pipe, the cold water supply to the hot water cylinder and the cold water supply to the bath, washbasin and WC. Pipes are measured in linear metres over all fittings, ignoring any joints (straight couplings) in the running length, which are deemed included. The location of the installation should also be provided. In most domestic situations the supply and service

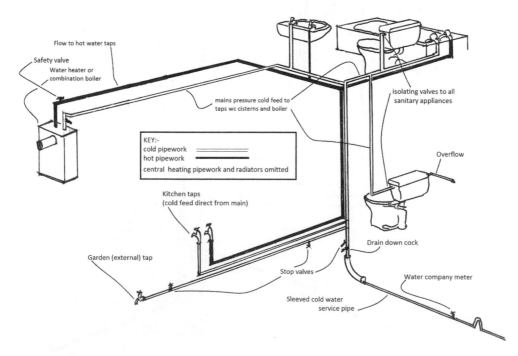

Figure 13.9 Direct system (mains-fed water).

pipes are likely to be plastic or copper, and this must be given in the description together with the pipe size, the method of jointing (compression or capillary) and the type, spacing and method of fixing. Curved pipework, fittings such as elbows, tees, reducers and tank connectors, together with made bends, are deemed included. Pipework in a roof space should be insulated in accordance with NRM2 38.9.2, as should the top and sides of the cold water storage cistern (NRM2 38.14). The measurement of tank stands, platforms and supports should also be included at this stage. Waste pipework, traps and overflow pipework are each measured in similar fashion and are considered further elsewhere.

13.04.04 Equipment

In a domestic situation this will include boilers, pumps, cisterns and cylinders. Work Section NRM2 38.1 or NRM38.2 provides the rules for the measurement of primary or terminal equipment. Both should be enumerated, giving in the description the type, size and pattern, rated duty, capacity, loading as appropriate and method of fixing. Since this could include a range of equipment intended to perform different functions, any description is perhaps best cross-referenced to a drawing or specification prepared by the services engineer.

The following sections describe the component systems associated with domestic plumbing installation, and are followed by a series of take-off lists for the same installation. There is no particular significance intended in the sequence of measurement, other than the need to establish a consistent approach.

13.05 Hot water installation

This will include the measurement of the boiler (NRM2 38.1), the hot water storage cylinder (NRM2 38.2) and the feed and expansion tank (NRM2 38.2). These should be enumerated, giving in the description the type, size and pattern, rated duty, capacity, loading as appropriate and method of fixing. Since this may include a range of equipment intended to perform different functions, any descriptions are perhaps best cross-referenced to a drawing or specification prepared by the services engineer. The associated hot water supply pipework would also be included, measured under the same set of rules as cold water pipework.

13.06 Heating installation

In a domestic situation where a gas-fired boiler is installed, the hot water and heating installation will share the same heat source. It would be possible to include the measurement of the boiler and hot water cylinder with either the heating or hot water installation using the same set of rules. Likewise, associated pipework should be measured as previously described. Radiators are enumerated as terminal pipeline equipment (NRM2 38.2.1) while radiator valves, when not supplied with the radiator, are enumerated separately (NRM2 38.5.1) as ancillaries to the radiator. Often a radiator schedule will be available, and descriptions can be extracted from this and presented under an appropriate heading, thereby preventing unnecessary repetition. Most radiators will be delivered to site pre-finished. Where this is not the case, an item for their decoration must be included in accordance with NRM2 29.4.1–3.1. The description should state the radiator type (panel or column) and it should be noted that the unit of measurement will vary depending on the area and girth of the radiator. Pumps, heating controls, motorised valves, immersion heaters and other control equipment associated with the heating system should be included in the same way. While these items can be measured at this point, the cabling and wiring of the control equipment can also be measured with the electrical installation. Care should therefore be taken to ensure that the wiring for this equipment is not included twice, or worse still omitted.

13.07 Sanitary appliances

This includes washbasins, pedestals, baths, bidets, shower trays, WC pans/cisterns and suites. These should be enumerated in accordance with NRM2 32.1 or NRM2 32.2 depending whether they are supplied with or without attached services, giving details in the description of the manufacturer's reference, the size and capacity, together with the method of fixing and the nature of the background to which they are fixed. Any ancillaries provided with these appliances, including supports and mountings, should be enumerated (NRM2 32.3.1). Frequently the cost of supplying sanitary appliances is included as a supply Cost Sum (see Chapter 4). Marking position, connecting to services and the provision of any supports are deemed included.

13.08 Waste pipework

The rules for the measurement of waste pipework are included in NRM2 Work Section 33, Drainage above ground level. A distinction is made between foul drainage installation above and below the ground, and the classification for each is given separately. Work below the ground is classified as NRM2 34 and considered further in the next chapter. Foul drainage pipework above the ground is often referred to as 'waste pipework', and this term is adopted here in the same context. It embraces all the pipework and fittings associated with the disposal of used or soiled water. In most modern installations, UPVC pipework will discharge into an internal downpipe, known as a stack or soil and vent pipe (S & V pipe); this in turn is connected to the drainage system. The following standard diameters of pipe are required for various appliances:

wash-hand basin – 32 mm diameter;
bath/shower/sink – 40 mm diameter;
WC – 110 mm diameter.

Overflow pipework (20 or 22 mm diameter) may well be treated and measured under the same set of rules, and will be required for WC cisterns, cold water storage cisterns and the feed and expansion cistern.

Waste pipework is measured in linear metres, giving in the description details of the type of pipe, its nominal size, whether it is straight, curved or flexible, and the spacing and type of pipe brackets. The background and method of fixing must also be stated, as should pipework that runs in ducts, chases, floor screeds or in situ concrete. Straight couplings in the running length of the pipes are deemed included, while elbows, tees, tank connectors and access plugs are enumerated as extra-over the pipework on which they occur. Where the diameter of the pipe fittings is less than 65 mm, the description must state the number of connections (NRM2 33.1–2.1–4). Where the diameter of pipe fittings exceeds 65 mm, the type of fitting must be stated. Some appliances, such as WC pans, have integral traps, while others (basins, baths and showers) require separate traps. When measured as separate items these are enumerated as pipework ancillaries in accordance with (NRM2 33.2.1), stating the type, method of fixing and nominal size. Testing and commissioning the waste pipework is measured as an item, giving details in the description of the tests and any attendance required (NRM2 33.10). In most cases the position of any sanitary appliance will be shown on the drawing. Similarly, the stack (ventilation pipe) and the position of the drainage connection should be shown. Given this information, it should be possible to measure the waste pipe runs, making appropriate allowances for vertical drops. The stack itself should be carried above the roof line to provide ventilation to the disposal system; alternatively a pressure release valve may be used in a stub stack installation.

13.09 Builders' work in connection

In the same ways as builders' work was included for electrical installation, so it should be measured for plumbing and heating installations. A general 'builders' work in connection item' would be recorded for each of the plumbing and heating installations. When measuring new work, this would mean a separate itemised builders' work in connection measurement for hot and cold water distribution, together with any central heating pipework and waste pipework. The same would also be required for any equipment and sanitary appliances. In each case a separate set of itemised measurement would be necessary for marking the position of holes, mortices and chases. Work for service installations to existing buildings is dealt with under a separate subset of rules (see NRM2 41.8 – 41.12).

13.10 Example take-off list

The approach to measuring service installation is very much a matter of individual choice. Some surveyors prefer to follow the flow of water as it enters and is distributed around the building, while others choose to measure the 'separate systems' as they are installed; as a consequence, no specific approach is prescribed. Whichever is adopted, dimensions should be recorded clearly and consistently. The following take-off list relates to no specific example and is intended to be typical of the items that are likely to occur in a low-rise residential detached property. It assumes a general approach as follows:

• Water service connection;
• Rising main;
• Cold water feed;
• Heating installation;
• Hot water feed;
• Sanitary appliances;
• Foul drainage above ground;
• Builders' work in connection;
• Testing and commissioning.

In addition to the take-off lists shown below, please see 13.11 for an example of typical booked dimensions.

Take-off list, NRM2 references
1 Water service connection

Connection to main prov. sum
Underground service runs: 41.13.1–5.1–3
Pipework (main supply pipe): 38.3.1.1
Pipe ancillaries (stop valve and drain tap): 38.5
Builder's work with
water main installation: 41.1.1
Stopcock pits: 41.22.1–3.1–7
Surface boxes : 41.21.1–3.1–7
2 Rising main
Rising main and branches (pipework): 38.3
Pipework ancillaries (pipe fittings): 38.4
Terminal equipment and fits (storage cistern): 38.2.1
Overflow to cistern (pipework): 38.3
Pipework ancillaries (pipe fittings): 38.4
Pipe insulation: 38.9.2.*.1
Cistern insulation: 38.9.1.1.1
Builders' work in connection with
Rising main installation: 41.1.1
Marking position of holes: 41.2
Testing and commissioning: 41.27
3 Cold feed
Pipework to taps (cold pipework): 38.3
Fittings to same (pipe fittings): 38.4
Pipe ancillaries (gate valves): 38.4
Pipe insulation: 38.9.2.*.1
Builders' work in connection
General BWIC cold feed installation: 41.1.1
Marking position of holes: 41.2
Testing and commissioning: 41.27
4 Heating installation
Primary equipment (boiler): 38.1.1.1
Pipework (inc. fittings): 38.3.1.1
Terminal equipment + fittings* (HW cylinder): 38.2.1.1.1
Terminal equipment + fittings (header tank): 38.2.1.1.1
Header tank insulation: 38.9.1.1.1
Immersion heater element (*included above)
Pipe ancillaries (motorised valves): 38.5
Pipe ancillaries (pump): 38.5
Pipe ancillaries (heating controls): 38.5
Terminal equipment + fittings (radiators): 38.2.1.1.1
Pipework (inc. fittings): 38.3.1.1
Pipe ancillaries (radiator valves): 38.5
Pipe insulation: 38.9.2.*.1
Builders' work in connection
General BWIC heating installation: 41.1.1
Painting pipes: 29.6.1.1
Painting radiators: 29.4.3.1

Marking position of holes: 41.2

Testing and commissioning: 41.27

5 Hot water feed

Pipework to taps (hot pipework): 38.3

Fittings to same (pipe fittings): 38.4

Pipe ancillaries (gate valves): 38.4

Pipe insulation: 38.9.2.*.1

Builders' work in connection

General BWIC hot water installation: 41.1.1

Marking position of holes: 41.2

Testing and commissioning: 41.27

6 Sanitary appliances

Fixtures, fittings with services: wash-hand basin, bath, WC + cistern, etc.: 32.2.3

Ancillary items (taps, waste outlets, etc.): 32.3

Builders' work in connection

General BWIC sanitary appliances: 41.1.1

Painting pipes: 29.6.1.1

Marking position of holes: 41.2

Testing and commissioning: 41.27

Take-off list, NRM2 references

7 Foul drainage above ground (wastes, overflows and traps)

Waste pipework (soil pipes/overflow pipes): 33.1.1–3.1

Pipework ancillaries (traps): 33.2

'Extra-over' for fittings to pipes: 33.3.1–2.1–4

Pipe sleeves: 33.4.1–3

Builders' work in connection

General BWIC waste pipe installation: 41.1.1

Marking position of holes: 41.2

Testing and commissioning: 41.27.

13.11 Examples of measured items for plumbing and electrical installations (Job reference P&E 1011/2007)

Sample plumbing and electrical installation take-off   Ref: P&E 1011/2007   Page 01 of 10

See NRM2 41 for accompanying drawings and minimum information required for measurement

Work for external service installation

15.00	Underground service runs average 1000mm deep, multiple pipe ducts supplied by others including:-
	38mm Ø electricity
	54mm Ø communication
	60mm Ø gas
	25mm Ø water

NRM2 41.13

Electrical Installation

1	Pipe duct fittings 'hockey stick' connected with plastic coupler and built into masonry work awp

NRM2 41.15.1.1

&

Accessories; recessed plastic meter cabinet 409mm wide x 595mm high x 210mm deep including lintel, polythene surround building in awp

NRM2 41.16.1.1

See NRM2 39 for accompanying drawings and mandatory information required for measurement

1	Primary equipment high level wall mounted 15 way dual split RCD consumer unit 63A RCDS c/w 12 MCB's 100A main switch, 2 x 63A 30mA RCD

NRM2 39.1.1.1

Sample plumbing and electrical installation take-off | Ref: P&E 1011/2007 | Page 02 of 10

<u>Power installation</u>

1 | Final circuits , cable and concealed containment in ground floor ring main consisting of 2.5mm² PVC insulated colour coded cable for 16 nr double switched recessed socket outlets . | NRM2 39.7.1

<u>Lighting installation</u>

1 | Final circuits , cable and concealed containment in first floor lighting circuit consisting of 1.5mm² PVC insulated colour coded cable for 28 nr ceiling mounted recessed light fittings. | NRM2 39.7.1

Item | General builder's work in connection with power and lighting installation | NRM2 41.1.1

&

Marking position of holes, mortices and chases | NRM2 41.2

&

Testing and Commissioning electrical installation all in accordance with BS 7671 Electrical Installation Certificate (EIC) standards | NRM2 41.27

Sample plumbing and electrical installation take-off   Ref: P&E 1011/2007   Page 3 of 10

See NRM2 38 for accompanying drawings and mandatory information required for measurement

Heating Installation

1	Primary equipment, heating system, Worcester/Bosch Greenstar 35CDi gas fired boiler, 34kW output complete with electrical controls, gas burner, control valve, automatic ignition boiler stat flue and flue terminal, wall mounting brackets and stove enamelled casing.	NRM2 38.1.1.1
1	Primary equipment, heating system, L1B compliant copper vented and insulated hot water cylinder 1500 x 300mm 90 litre, pre-holed for immersion element and 3nr 22mm Ø connections	NRM2 38.1.1.1
1	Terminal equipment and fittings, heating distribution system to ground and first floor with Myson Select Compact DC pressed steel radiators to BS EN 442 and BS EN ISO 9001 complete with thermostatic radiator valves with 15mm angled connections; radiator mounting brackets fixed to masonry walls	NRM2 38.2.1.1.1

Sample plumbing and electrical installation take-off   Ref: P&E 1011/2007   Page 04 of 10

<u>Heating Installation (Contd)</u>

Item — General builder's work in connection with heating installation   NRM2 41.1.1

&

Marking position of holes, mortices and chases   NRM2 41.2

&

Testing and Commissioning heating and gas installation by Gas Safe registered engineer.   NRM2 41.27

<u>Cold water distribution installation</u>

<u>Copper tube to BS EN 1057 with end feed capillary fittings</u>

4/ 6.50
2.80
3.20
0.90
4.65
2/ 3.60
2.40
0.70

Pipework 15mm Ø to ground, first floor and roof space in cold water distribution system   NRM2 38.3.1.1

Note all fittings deemed included

300 *Drainage, electrics and plumbing*

Sample plumbing and electrical installation take-off   Ref: P&E 1011/2007   Page 05 of 10

	Cold water distribution Installation (contd)	
1	Pipework ancillaries, brass high pressure gate valve with 2nr 15mm Ø connections to copper tube	NRM2 38.5.1
1	Primary equipment, cold water distribution system, Ferham enclosed range ref FCGR50 size 965 x 635 x 610mm including insulation jacket 60mm GRJ50 and water regulation kit GK50, holed for one 15mm Ø and three 22mm Ø connections	NRM2 41.27
1	Pipework ancillaries, brass high pressure ball valve with plastic float assembly and connection to copper pipe	NRM2 38.5.1
	Solvent welded PVC overflow pipework	NRM2 38.3.1.1
3.80 0.75 0.90	Pipework 15mm Ø to roof space, first and ground floor ground, in cold water distribution system	Note all fittings deemed included

Sample plumbing and electrical installation take-off   Ref: P&E 1011/2007   Page 06 of 10

<u>Cold water distribution Installation (Contd)</u>

Item	General builder's work in connection with cold water distribution installation	NRM2 41.1.1

&

Marking position of holes, mortices and chases   NRM2 41.2

&

Testing and Commissioning cold water installation.   NRM2 41.27

See NRM2 32 for accompanying drawings and mandatory information required for measurement

<u>Sanitary appliances</u>

1   Fixture, fittings or equipment with services, vitreous china Ideal Standard Alto close coupled cistern 6/4 litre dual flush valve bottom supply and internal overflow with matching seat and cover 395mm wide x 685 mm projection x 810mm high, screw fixing cistern to masonry and pan to timber floor   NRM 2 32.2.1-3.8.1-2

Sanitary appliances (Contd)

1	Fixture, fittings or equipment with services , Ideal Standard vitreous china Alto 550 washbasin, one tap hole with overflow complete with matching full pedestal including brass mixer tap waste fitting, plug chain and stay, 550mm wide x 450mm projection x 855mm high , plug and screw wall brackets to masonry and pedestal to timber floor	NRM 2 32.2.1-3.8.1-2

Sanitary Appliance Installation

Item	General builder's work in connection with sanitary installation	NRM2 41.1.1
	&	
	Marking position of holes, mortices and chases	NRM2 41.2
	&	
	Testing and Commissioning sanitary installation.	NRM2 41.27

Sample plumbing and electrical installation take-off   Ref: P&E 1011/2007   Page 8 of 10

See NRM2 33 for accompanying drawings and mandatory information required for measurement

<u>Drainage above ground</u>

6.50	PVC-U pipework 110mm Ø		
1.80	straight ring seal joints	NRM2 33.1.1.2	
2.40	Osma reference 4S044G		
	including screw fixing pipe		
	brackets to masonry		
	backgrounds		

Extra over ditto for:-   NRM2 33.3.2.1

1   Fittings nominal pipe size >
65mm Ø one end

NRM2 33.3.2.2

5   Fittings nominal pipe size >
65mm Ø two ends

NRM2 33.3.2.3

3   Fittings nominal pipe size >
65mm Ø three ends

1   PVC-U pipework ancillaries   NRM2 33.2.1
110mm Ø roof and terminal
fittings (balloon grating)
reference 4S302G

&   NRM2 33.2.1

PVC-U pipework ancillaries
Multikwik pan connector
reference MKB21104

&   NRM2 33.2.1
or
NRM2 18.4.1.8

PVC-U pipework ancillaries
roof and terminal fittings,
aluminium pipe flashing 460
x 460mm reference 4S283G

Sample plumbing and electrical installation take-off   Ref: P&E 1011/2007   Page 09 of 10

Drainage above ground (Contd)

1.80 0.90 1.40	PVC-U pipework 32mm Ø straight ring seal joints Osma reference 4M073E including screw fixing pipe brackets to masonry backgrounds	NRM2 33.1.1.1/3.1
	Extra over ditto for:-	
4	Fittings nominal pipe size ≤ 65mm Ø one end	NRM2 33.3.1.1
6	Fittings nominal pipe size ≤ 65mm Ø two ends	NRM2 33.3.1.2
2.10 0.90 0.60	PVC-U pipework 40mm Ø reference 5Z074G including screw fixing pipe brackets to masonry backgrounds	NRM2 33.1.1.1/3.1
	Extra over ditto for:-	
2	Fittings nominal pipe size ≤ 65mm Ø one end	NRM2 33.3.1.1
4	Fittings nominal pipe size ≤ 65mm Ø two ends	NRM2 33.3.1.2

Sample plumbing and electrical installation take-off   *Ref: P&E 1011/2007   Page 10 of 10*

Drainage above ground (Contd)

3	PVC-U pipework ancillaries 32mm Ø Osma TPP anti-syphon bottle trap WT 32 reference 4V814W	NRM2 33.2.1
2	PVC-U pipework ancillaries 40mm Ø Osma TPP anti-syphon bottle trap WT 40 reference 5V814W	NRM2 33.2.1
Item	General builder's work in connection with drainage above ground	NRM2 41.1.1
	&	
	Marking position of holes, mortices and chases	NRM2 41.2
	&	
	Testing and Commissioning sanitary installation.	NRM2 41.27

14 Disposal systems below ground

14.01 Introduction

Drainage below ground provides for the dispersal of foul and storm water from a building to a point of disposal or treatment. The New Rules of Measurement 2 distinguish between disposal systems above and below ground level (NRM2.33 and NRM2.34, respectively). An explanation of the approach to measuring disposal systems above ground is provided in Chapter 13 (see Waste pipework, 13.08). Any drainage work below existing buildings and any work outside the boundary of the site will need to be identified and measured separately. As with other NRM2 work sections, it will be necessary to provide details of the proposed drainage system layout in support of the tender documents. This information, together with details of all works and materials that are deemed included, can be found in the opening section of NRM2 34. For any estimators not familiar with NRM2, it would be sensible to check the 'deemed included' list since this now includes many items that in previous versions of the Standard Method of Measurement were measured and costed as separate items.

14.02 General sequence of measurement

Assuming the group method of measurement is adopted, disposal systems above the ground would be measured with their associated work (i.e. rainwater goods with roof coverings

(Chapter 10) and waste pipework with plumbing installations (Chapter 13)). NRM2 Work Section 34 provides the classification and measurement rules for the following below ground drainage installation:

- Storm water systems;
- Foul drainage systems;
- Pumped drainage systems;
- Land drainage.

Each will require the excavation of trenches, the laying of pipe runs, the provision of beds and the construction of manholes or inspection chambers. In addition, a pumped drainage system will inevitably require the provision of specialist drainage pumping equipment (NRM2 34.5). For the purposes of measurement, it is appropriate to consider the provision of manholes separately from pipe runs. Where separate systems for foul and storm water are proposed, it is sensible to measure each system independently.

The line of drainage runs, together with the position of manholes, will be shown on the drawings. In addition, the drawings should identify each manhole by a reference number (MH1, MH2 etc.) and give details of the cover and invert level. If no reference is given, before measurement can commence it will be necessary to commence at the top of each drainage run and identify each manhole with a specific reference. This is vital since it will provide a unique location code for both schedules and take-off. Where existing and proposed ground levels are provided to coincide with the location of manholes, all to the good; where this is not the case, these will need to be established by interpolation if necessary (see Chapter 5.02.03, Cut and fill).

A consistent, logical and fully annotated set of dimensions is necessary for this class of work, and the preparation of schedules for all but the smallest of systems is recommended. Some surveyors prefer to measure manholes first, while others choose to measure drainage runs; either approach is acceptable. The following sequence of measurement has been adopted in the preparation of this text.

Foul drainage

1 Main drainage runs;
2 Branch drainage runs;
3 Pipe fittings and accessories;
4 Manholes, soakaways, septic tanks, inspection chambers, cesspits;
5 Sewer connections;
6 Testing and commissioning.

Surface water and land drainage

1–3 as above;
4 Soakaways;
5 Testing and commissioning.

14.03 Foul drainage

14.03.01 Main drainage runs

These are measured in linear metres on plan and are based on a conglomerate approach to measurement where excavation, bedding materials and drainage pipe (with associated connecting sleeves) are all included in a single, all-embracing measurement. Earthwork support, compacting the bottom of trenches, trimming, backfilling, disposal of excavated materials and disposal of water are all deemed included. At this stage it is possible to ignore any pipe fittings (bends, junctions, rodding eyes and the like) as these can be 'picked up' later by measuring 'extra-over' the pipework on which they occur. Since pipes are built into the sides of manholes and inspection chambers, the length of the excavation and the length of the pipe may differ. For the purposes of measurement, this minor discrepancy is now deemed included. For a full list of the items 'deemed included', reference should be made to the opening section of NRM2, Work Section 34.

14.03.02 Manhole and pipe run schedules

Before commencing the take-off for drainage installation, the preparation of both a pipe run and manhole schedule is recommended. While this might be considered unnecessary for small-scale work to an existing drainage system, for anything bigger, the preparation of both pipe run and manhole schedules is recommended. It is advantageous to prepare the manhole schedule before commencing the pipe run schedule, since some of the information for pipe runs can be recycled from the manhole schedule (e.g. the depths of pipes will be determined by the invert depths of the manholes). Typically schedules are prepared on a sheet of abstract paper with the following headings (see Figures 14.1 and 14.2). These are intended as a guide only and should be amended to suit the scale and scope of the drainage installation being measured.

14.03.03 Drain runs

These are measured in linear metres, stating the average trench depth in increments of 500 mm; the same description should include the type and nominal diameter of the pipe. Where

M/H No.	Internal Size		Ground Level	Inver. Level	Depth to I.L.	BKWK Sides	Slab or Cover	Main Chann.	Conc. Bench	Foot Irons	Remarks
	Length	Width									
1	1.350	0.778	19.500	17.650	1.850	215 ff o/s	600 × 450 light duty c.i. on 100 mm pcc slab	100 mm ϕ st.	nil	5 no	All brickwork in class 'B' bricks
2	0.778	0.553	16.750	15.963	0.787	ditto	ditto	100 mm ϕ curved	1−100 mm ϕ	nil	ditto
3	0.675	0.450	17.050	16.375	0.675		ditto	100 mm ϕ st.	2 −100 mm ϕ	nil	ditto
4	0.675	0.450	17.250	16.650	16.600	ditto	ditto	ditto	1−100 mm ϕ 3/4 sect	nil	ditto

Figure 14.1 Typical headings for manhole schedule.

Loc	Length between inside faces of MH	Pipes			Length between outside faces of MH	Concrete beds			Length between exc. faces of MH	Depth of dig one end	Depth of dig other end	Av. depth of dig
		Type	Size	Bends		Bed	Haunch	Sur				

Figure 14.2 Typical headings for pipe run schedule.

multiple pipes are laid in the same trench, the number of pipes and their nominal diameters should be included in the description. Details of the method of jointing pipes, the type of bedding and pipe surround material, together with the type of backfill (assuming that it is not backfill materials obtained from the same excavation), must also be included in the description. Finally, as and where necessary, the description should identify any vertical drainage runs, drain runs next to existing buildings, and drain runs that are laid below groundwater, to name only three of the seven specific conditions detailed as requirements in NRM2 34.1*.*.1–7 (Figure 14.3).

As mentioned in the opening paragraph of this chapter, in addition to the pipe (together with any bedding material and pipe covering), the pipe run trench excavation is deemed to include earthwork support, compacting trench bottoms, trimming excavations, filling with and compaction of filling materials and the disposal of any surplus excavated material.

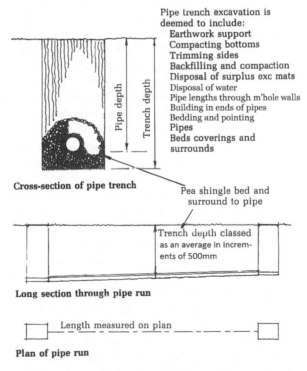

Figure 14.3 Plan, longitudinal and cross-section view of drainage pipe runs showing the measurement of drainage pipe runs in linear metres giving average depths in increments of 500mm.

Where it is necessary to break out existing hard surface paving or to remove turf, this should be measured 'extra-over' the previous trench excavation, stating the thickness of the material to be removed and given in square metres. Once this work is complete, details of any re-instatement (or relaying of turf) will be covered by an inclusion to this effect in the description. Breaking out any hard materials (presumably those met when excavating the trench), together with removing hazardous materials, excavating in unstable ground (running silt) or below groundwater level, are all given in cubic metres (NRM2 34.2.3,4,5,6). Separate special consideration is given when excavating drain runs next to or around existing live services. In the first instances (next to), these are measured in linear metres (NRM2 34.2.7) and where any live services cross the trench, these are measured as enumerated items (NRM2 34.2.8).

14.03.04 Manholes and inspection chambers

The terms 'inspection chamber' and 'manhole' are used interchangeably in this text, although it is recognised that there is a distinction between them. The term 'inspection chamber' allows for a visual inspection at the point of access to the drainage system, while 'manhole' implies that in addition to a visual inspection, it is possible for a person (not exclusively a man) to access the drainage system for maintenance purposes (Figure 14.4).

It is suggested that the preparation of an inspection chamber or manhole schedule should precede the take-off (see 14.01). Increasingly in domestic situations, preformed plastic inspection chambers are used (Figure 14.5). When compared to brick-built or preformed concrete section inspection chambers, these are easier and cheaper to install; however, they are usually only permitted for inspection purposes where the depth of the chamber is less than 1.20 m (or in some cases 3.00 m). Where the depth of the inspection chamber exceeds 1.20 m, either a traditional brick or precast concrete chamber would be expected (Figure 14.4).

When measuring manholes, inspection chambers and the like, these are enumerated giving a description of the type of chamber including the maximum internal size and depth from top surface of cover to invert level (in 250 mm stages). When measuring proprietary systems, details of the system must be stated. The associated work of excavation, base slab, cover slab, benching, main channel diameter and configuration, together with the number and diameter of any branch channels, should all be given in the description (NRM2 34.6–11.*.1–9). Sundry items associated with inspection chambers, including step irons, intercepting traps, backdrops etc., are enumerated (NRM2 34.13.1.1–4). Covers and frames and marker posts are also enumerated (NRM2 34.14.1.1.1.1 and NRM2 34.15.1.1.1.1). Note should be made of the different treatment when these items form part of a preformed system.

Where it is necessary to break out existing hard surface paving or remove turf, these are measured 'extra-over' the previous inspection chamber pit excavation in square metres, stating the thickness of the hard paving material to be removed. Once the work is complete, details of any reinstatement (or relaying of turf) can also be included in the description. Breaking out any hard materials (presumably those met when excavating the inspection chamber pit), together with removing hazardous materials, excavating in unstable ground (running silt) or below groundwater level, are all recorded in the dimension column in cubic metres (NRM2 34.12.3,4,5,6).

TYPICAL RECTANGULAR
PRECAST CONCRETE INSPECTION
CHAMBER

LONG SECTION

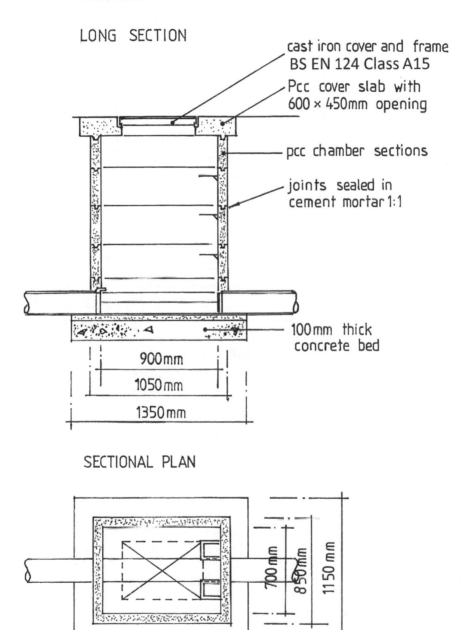

cast iron cover and frame
BS EN 124 Class A15

Pcc cover slab with
600 × 450mm opening

pcc chamber sections

joints sealed in
cement mortar 1:1

100mm thick
concrete bed

900mm

1050mm

1350mm

SECTIONAL PLAN

700 mm

850 mm

1150 mm

Figure 14.4 Rectangular precast concrete inspection chamber.

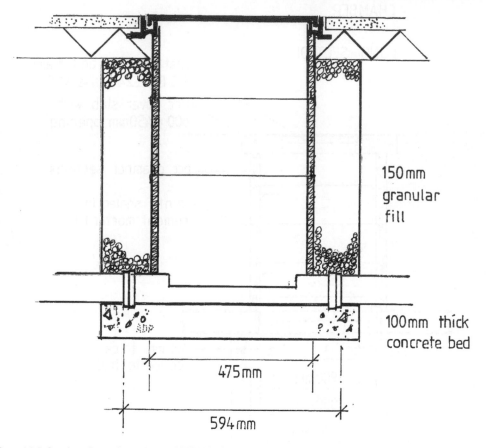

Figure 14.5 Section through preformed UPVC circular inspection chamber (450mm internal diameter).

14.03.05 Connection to mains sewer

In the normal course of events, the connection to the main sewer will be carried out by the local water supply/waste water treatment company. Where this is the case, the cost of this work can be included by way of a Defined Provisional Sum in accordance with NRM2 2.9.4.1. Since it is likely that this will involve the breaking up and reinstatement of a public highway, the inclusion of a separate Defined Provisional Sum to cover these further costs may also be required.* This work may be incorporated into the BQ in a number of ways, but provision is made in NRM2 34 (Drainage Below Ground) for this to be included as an item under the heading of 'Connections' (NRM2 34.16).

* For trench excavation beyond the boundary of the site, breaking up and later reinstating paved surfaces and any special provision for traffic control and pedestrian safety costs need to be included.

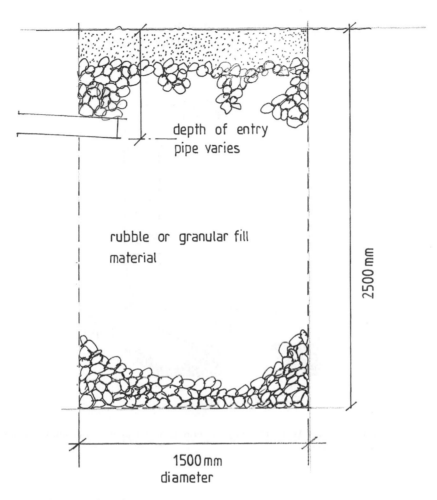

Figure 14.6 Section through soakaway.

A booked dimension by way of an item must also be made for the testing and commissioning of the drainage system before it can be adopted. This should provide details of the type of test and the standard to be achieved, including the method, purpose and type of installation to be tested. Where attendance or any special instrumentation is required, this should also be given in the description, as should preparatory operations and any stage tests, giving the number of tests (nr) in the description.

14.04 Soakaways

Where it is not permissible or appropriate to discharge rainwater into a sewer, soakaways should be provided. There are specific rules about their construction and location. In domestic situations where ground conditions allow, they are constructed as simple pits filled with granular material (Figure 14.6).

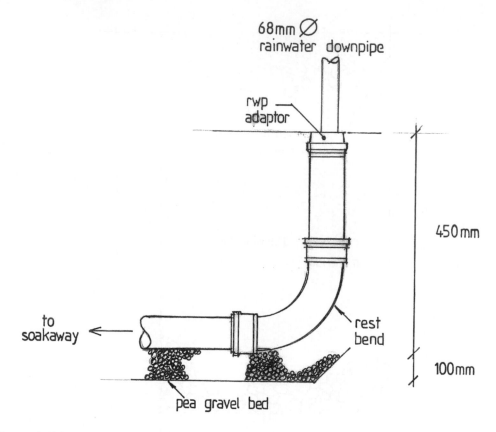

Figure 14.7 Below Ground, Surface Water Drainage; Section through rainwater downpipe above to below ground connector.

The measurement of excavation items, together with the pipe runs associated with soak aways, should be carried out using the same set of rules adopted for other drainage work. No specific mention is made in NRM2 with regard to filling of soakaway pits. In the absence of any specific guidance, it is assumed that this would be given in accordance with Excavating & Filling (NRM2 5.12.3) in cubic metres as imported filling, beds exceeding 500 mm deep. The drainage below ground level example take-off that follows has been prepared based on this approach. Details of the connection between above ground rainwater disposal and below ground discharge drainage pipework are shown in Figure 14.7.

Wherever possible, consideration should be given to rainwater harvesting and/or sustainable drainage systems (SUDS) as an alternative to a soakaway. An example of measurement for the latter is provided in Chapter 15, External works.

14.05 Drainage below ground drawing and specification

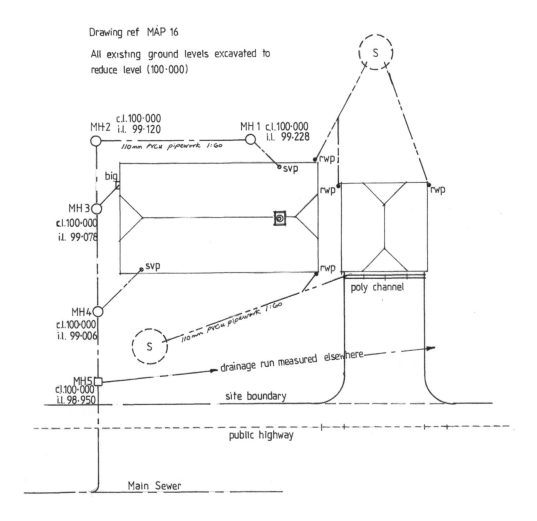

Drawing ref MAP 16

All existing ground levels excavated to
reduce level (100·000)

Drainage key:-

○ Preformed PVCu inspection chamber

▢ Precast concrete sectional manhole

(S) Soakaway

svp Soil and vent pipe connection

big Back inlet gully

rwp Rainwater pipe connection

c.l. Cover level

i.l. Invert level

——— — ——— foul drainage runs

——— — — —— surface water drainage runs

All pipework 110 mm ⌀ PVCu laid to falls 1:60

Pea shingle bed and surround to foul pipes, bed only to s.w. pipes

Provisonal Sum inclusion for work beyond site boundary

14.06 Drainage below ground, measurement (Job ref: MAP 16)

Drainage below ground Detached dwellings with Detached Garage Job ref: MAP 16 ; Page 01 of 07

Mandatory Information: The scope of the work involves the provision of foul and surface water drainage to a pair of detached dwellings with a shared detached garage . Layout, invert levels, cover levels and pipe sizes all in accordance with drawing MAP 16 and accompanying specification notes and drainage /inspection chamber schedule

Take-off list

FOUL DRAINAGE SYSTEMS

- main drainage runs
- branch drainage runs
- pipe fittings
- accessories
- inspection chambers /mholes
- associated sundry items
- covers and frames
- testing and commissioning

Work beyond the boundary of the site
- break up hard surfaces and reinstate
- main drainage runs
- connections
- provisional sum
End of work beyond site boundary

Needs to be measured separately
See NRM2 34 Notes

STORM WATER DRAINAGE
SYSTEMS

- main drainage runs
- branch drainage runs
- pipe fittings
- accessories
- soakaways
- testing and commissioning

Note:- Given the nature of the substrata the Structural Engineer has stipulated that temporary support to the sides and faces of open excavation will not be required .

It is assumed that the site has been cleared of all vegetation and clear contractor access is possible.

See NRM2 34 for accompanying drawings and minimum information required for measurement

Drainage below ground – Detached dwellings with Detached Garage; Job ref: MAP 16 ; Page 02 of 07

FOUL DRAINAGE - MAIN RUNS

Average excavation depth of
pipe runs (see MH schedule)

Top of cover to invert level
   MH 1   872mm
   MH 2   980mm
average depth = _500 – 1000mm_

   MH3   1022mm
   MH4   1094mm
   MH5   1175mm
average depth = _1000 - 1500mm_

6.50	
2.50	

Drainage runs average (MH 1-2)   NRM2 34.1.1/2.1-3
trench depth 500 - 1000mm (MH2-3)
PVCU pipes BSEN 1401-1
110mm Ø with ring seal sockets,
bedded in pea gravel beds and
surround.

4.30
2.80
1.00

Ditto average trench (MH 3-4)   NRM2 34.1.1/2.1-3
depth 1000mm – 1500mm (MH 4-5)
ditto (MH 5 to site boundary)

FOUL DRAINAGE
BRANCH CONNECTIONS

1.50
1.20

Drain runs trench av (MH 1)   NRM2 34.1.1/2.1-3
depth 500 - 1000mm (MH 3)
as before described

2.50

Drain runs trench av (MH 4)   NRM2 34.1.1/2.1-3
depth 1000 - 1500mm
as before described

318   *Disposal systems below ground*

FOUL DRAINAGE
BRANCH CONNECTIONS
(Contd)

3 / 0.90	Drain runs trench av depth 500 - 1000mm all as before described but vertical work inc. pea gravel surround and bed (svp x 2) (big x 1)	NRM2 34.1.1/2.1-3.2

svp = soil and vent pipe
big = back inlet gully

PIPE FITTINGS

2 / 1	Pipe fittings for UPVC pipes Osma long radius Rest bend 110 mm Ø ref 4D181 including ring seal Connector bedded and surrounded in peas gravel (svp x 2)	NRM2 34.3.1.1/2

ACCESSORIES

1	Accessories UPVC Osma bottle gulley ref 4D900 including ring seal connector and pea gravel surround (BIG)	NRM2 34.4.1.1/2

END OF FOUL DRAINAGE

The following foul drainage
system is outside the boundary
of the site

2.50	Drain runs trench av. depth 1000 - 1500mm as before described (site boundary to mains)	NRM2 34.1/2.1-3

Drainage below ground – Detached dwellings with Detached Garage; Job ref: MAP 16 ; Page 04 of 07

	The following foul drainage system is outside the boundary of the site (Contd).	
2.50 0.45	Extra over drain run for (site boundary to mains) breaking up hard surface pavings 100mm thick and reinstating with 50mm thick hardcore base and 50mm macadam topping	NRM2 34.2.1.1
	DEFINED PROVISIONAL SUM	
£ def provis Sum	Include the defined Provisional Sum of £1,500.00 for work to provide mains sewer connections including reinstating hard pavings on completion	NRM2 Part 2 9.4.1
Item	Allow for overheads on last & Allow for profit (xx %) on last	NRM2 Part 2 9.4.1
	End of foul drainage system work outside of the site boundary	

<u>INSPECTION CHAMBERS</u>
<u>& MANHOLES</u>

<u>Osma UPVC underground</u>
<u>Universsal inspection chamber.</u>
<u>450mm internal Ø to</u>
<u>BSEN13598 part 1</u>

| 4/ 1 | Proprietary systems   (MH1 - 4) Inspection chambers comprising chamber base Osma ref 6D929 chamber shafts Osma ref 4D975, cover and frame Osma ref4D927 depth range top of cover to invert 750 -1000mm | NRM2 34.7.3.1-9 |

<u>Milton pre-cast concrete</u>
<u>Inspection chamber sections</u>
<u>BSEN 5911 Part 200. 900 x</u>
<u>700mm on plan with cast-in step</u>
<u>irons and lifting lugs; joints</u>
<u>sealed in cement motar (1:1)</u>

| 1 | Proprietary systems comprising pre cast concrete chamber section on in-situ concrete (1:2:4) 100mm thick base with 110mm Ø branch (all in UPVC pipework to match rest of system) 1100 x 850mm pre cast concrete cover slab with 600 x 450mm opening and light duty cover and frame to BSEN124 depth; top of cover to invert 1000 – 1500mm | NRM2 34.6.3.1-9 |

STORM WATER
                INSTALLATION

Average depth of pipe runs

                    head = 450
                    bed = 100
                              550

(worst case) 9 metre run
@ 1:60 fall = 150mm

therefore depth is 550
                            + 150
                               700

Average depth range 500 -1000

9.00	
5.80	
5.00	
1.00	
3.00	

Storm water drain (main runs)   NRM2 34.1.1/2.1-2
runs average trench depth
500 - 1000mm PVCU
pipes BSEN 1401-1 110mm Ø with
ring seal sockets, bedded (branch)
in pea gravel beds and surround.

4 / 0.45

Ditto but vertical work (rwp connex)   NRM2 34.1.1/2.1-3.2

PIPE FITTINGS

4 / 1

Pipe fittings for UPVC (rwp)
pipes Osma long radius rest   NRM2 34.3.1.1/2
bend ref 4D181 including ring
seal Connector for 110 mm Ø
outlet pipe and 68mm Ø rainwater
inlet pipe; pea gravel bed and
surround

PIPE FITTINGS

| 2/1 | Pipe fittings equal   Y branch   Junction 87.50 Osma ref 4D210 | NRM2 34.3.1.1/2 |

ACCESSORIES

| 1 | Accessories, precast concrete Polychannel SK 100 wide x 150mm deep box section with built in falls, including removable metal grating 5400mm long | NRM2 34.4.1.1/2 |

| 1 | Ditto SK100 sump box with lift-out bucket and 110mm Ø outlet and connection to UPVC storm water drainage system | NRM2 34.4.1.1/2 |

SOAKAWAY

| 2/1 | Form circular soakaway radius 750mm x 2500mm deep; imported brick rubble fill, galvanised metal sheet cover with topsoil backfill; build-in end of 110mm Ø UPVC pipe 2250 -2500mm deep | NRM2 34.8.1/2 |

| Item | Testing and commissioning surface water drainage system | NRM2 34.17.1 |

15 External works

15.01 Introduction

By way of general definition, external works can be described as works that form part of a construction/building project that are external to the main building structure. There are three separate work sections included in NRM2 that have been drafted specifically for items that would typically be associated with external works: NRM2 35 Site Works, NRM2 36 Fencing and NRM2 37 Soft Landscaping. In addition, reference is made in these three sections to NRM2 5 Excavating & Filling and NRM2 11 In situ Concrete Works. Even though it is not specifically mentioned, there is the possibility that the measurer may need to need to defer to NRM2 work section 6 Ground Remediation and Soil Stabilisation.

Examples of main item coverage for the work sections NRM2 35 Site Works, NRM2 36 Fencing and NRM2 37 Soft Landscaping are given below.

- (NRM2 35 Site Works) Permanent site works: in situ concrete, macadam and asphalt, roads, paths, paving, sheet linings to pools, ponds and waterways, site furniture, proprietary sports surfacing, tufted surfacing, line and road marking.
- (NRM2 36 Fencing) Fencing: close-boarded, palisade, chain link, metal welded security fencing and associated gates and gateposts.
- (NRM2 37 Soft Landscaping) Soft landscaping: cultivating, treating, soiling, seeding and turfing, trees, shrubs, hedge plants, protection and maintenance.

Although NRM2 has no specific work section titled 'External Works', all of the above are likely to be grouped together and included under this as a general heading in the completed BQ.

The earlier definition of the term 'external works' should be treated with some care and an element of 'local interpretation' applied. For instance, the boundary between what is considered 'internal' and 'external' could require some interrogation, particularly where projects intentionally blur internal and external features. Partly enclosed entrance porticos and atria to office or retail projects are examples of projects where it may be necessary to provide some further detail in order to make it clear to all involved what should be measured and costed where.

It should also be noted that there may be occasions when it would be convenient to measure and bill separately walls that are not associated with the fabric and structure of the main construction (external works walling). In these cases the measurer should defer to NRM2 Work Section 14 Masonry. Even though it is not specifically mentioned, it is suggested that for the purpose of billing, 'external works walling' is included with NRM2 35 Site Works under an appropriate heading.

15.02 Preparation for measuring external works

When measuring external works, the measurer is advised to familiarise themselves with the drawings, noting in particular existing and proposed ground levels, the presence of any hazardous materials and any features that are to be retained. Any works related to an environmental survey for the project are likely to have been carried out prior to the main contractor appointment, and it would follow that any costs associated are likely to have been allocated as a preliminary (works package contract) item (see control and protection, surveys, environmental monitoring (NRM2 1.4.1.5)).

15.03 Booking dimensions for external works

This includes topsoil, Surface Excavation, Filling, Disposal and Cut and Fill (Refer to Chapter 5, Substructures).

Some understanding of the data available from the site survey, including the grid of levels, will be necessary before it is possible to measure any surface excavation. Details of this and other associated excavation and filling measurement techniques (surface excavation, topsoil, bulk excavation, disposal and cut and fill) are explained and demonstrated in Chapter 5, Substructures. It is suggested that the reader reviews that chapter before attempting to measure any items associated with NRM2 35 Site works or NRM2 37 Soft landscaping. Particular attention should be given to establishing the existing ground level of any site. To this end, the following is offered to supplement the examples and figures given in Chapter 5.

15.03.01 Grid of levels

The best way to explain how a grid of levels works is to imagine a series of 10 × 10 m square grids set out on the surface of a site. Let us assume that the site in question is 40 × 20 m on plan and that the 10 × 10 m grids are arranged next to each other, thereby creating the rectangular plan shape we require. This would give four boxes along the length and two for the width (see Chapter 5, Figure 5.3). A ground-level reading is taken at each intersection of this grid (so in this case we will have a total of 15 readings). Each of these grid squares has four readings, one in each corner. By establishing the average ground level of each individual grid and then subsequently calculating the average of each of these grids (in other words, the average of all the averages), this will give the average existing ground level for the 40 × 20 m site. There is a more efficient way to achieve this same outcome, and this is known as the 'weighting method' (see Chapter 5, Figure 5.5).

15.04 Site works

This work section includes the rules for measuring roads, pavings, hard landscaping and sports surfacing (NRM2 work section 35).

In the context of building projects, road and paving construction is likely to be limited to the installation of estate and access roads. In normal circumstances, upon completion of the project, the local authority will 'adopt' the road construction. As such it is the design engineer's (or contractor's) responsibility to ensure that the road is designed and built to the local authority's specification.

It is likely that the measurement of roads and pavings will initially involve excavation, filling and disposal of excavated materials, but this will only be determined once existing and proposed finished road levels are known. There is no need for a separate set of measurement rules for this section of the work, since we are able to adopt the same as those designed to measure substructures for general building operations (NRM2 work section 5 Excavation and Filling).

Anyone involved with the measurement of external works will need to be familiar with recording dimensions for irregular areas, particularly those formed by road surfaces such as 'T'-junctions and cul-de-sac turning heads. For further guidance on this, reference should be made to Chapter 3 (sections 3.06 to 3.09 and Figures 3.15 to 3.20). In addition, the reader should review the examples of how to record dimensions based on the different geometric formulae as given in Appendix II, the online measurement guide and the take-off example that accompanies this chapter.

The following measurable items assume a greenfield site where preliminary site work has already been carried out and the provision of a flexible paving road surface is intended.

- Site preparation
  - lifting turf, NRM2 5.5.1.1
  - removing topsoil (and disposal), NRM2 5.2/5.10.1.1
  - removing any existing hard surfaces, NRM2 5.5.3.1
- Bulk excavation
  - reduce and surface excavation (and disposal), NRM2 5.6.1/5.9.2.1
- Kerb construction
  - kerb trench excavation (and disposal), NRM2 5.6.2.1/5.9.2.1
  - in situ concrete foundation, NRM2 11.2.1.2.1
  - precast concrete kerbs, NRM2 35.1.1.*1/35.1.1.1.1
  - precast concrete channels, NRM2 35.1.1.*1/35.1.1.1.1
- Road sub-base
  - imported filling, NRM2 5.12.2.1
- Road base course
  - dense bitumen macadam, NRM2 35.12.1.1.2

- Road wearing course
  - coated macadam, NRM2 35.12.1.1.2
- Footpath construction
  - kerb trench excavation (and disposal), NRM2 5.6.2.1/5.9.2.1
  - in situ concrete foundation, NRM2 11.2.1.2.1
  - precast concrete edging, NRM2 35.2.1.*1/35.2.1.1.1
- Footpath sub-base
  - imported filling, NRM2 5.12.2.1
- Footpath base course
  - dense bitumen macadam, NRM2 35.12.1.1.2
- Footpath wearing course
  - coated macadam, NRM2 35.12.1.1.2.

15.05 Fencing

Measured in linear metres, fencing (NRM2 36.1.1.1–4) is measured over all supports (fenceposts), giving the height of the fence panels, together with their spacing, height and the depth of the supports (fenceposts) in the description. Work set to a curve (but straight between posts), along with curved work, work where the ground slopes less than 15° or where individual fence panels are in excess of 3.00 m in height will all have to be so described. Special supports (end posts, angle or corner posts, integral gateposts, straining posts and the like) are all enumerated as 'extra-over' the fencing on which they occur. The provision of any independent gateposts is measured as enumerated items (nr), as are gates and any associated ironmongery.

The following items are deemed included (unless stated otherwise): excavation, backfilling and disposal of surplus materials, earthwork support, disposal of ground and surface water, in situ concrete for post filling, formwork and any temporary support.

Where the presence of groundwater is encountered in any excavation work associated with external works, this should be described as an 'extra-over' item and recorded as a volume (see conditions for pre- and post-contract groundwater level adjustments, Chapter 5.04). Breaking out existing 'hard' materials met in the excavation of fence supports is also described as an 'extra-over' item and given as a volume. A similar 'extra-over' approach is adopted for breaking up any existing hard surface pavings, but in this instance these are recorded in square metres, stating the thickness in the description.

15.06 Soft landscaping

This work section (NRM2 37) provides the rules for measuring the cultivation and preparation of the surface of the ground to receive seeding, turfing, the planting of trees, shrubs, bulbs and the like, together with their subsequent protection and maintenance. A relatively uncomplicated set of measurement rules is used, adopting either area (m^2), enumeration (nr) or item as the principle units of measurement.

The moving and placing of topsoil in and around the site should be measured and billed in accordance with NRM2 5 Excavation and Filling. The exact details (and NRM2 reference) will depend on whether the topsoil has been excavated previously and retained on site for later reuse (NRM2 10.1.1/2.1.1) or imported (NRM2 12.1–3.1–3.1).

Where contaminated ground is identified or hazardous materials encountered, this will require reference to NRM2 Work Section 6, Ground Remediation & Soil Stabilisation. An explanation of how this class of work should be measured is included elsewhere (see Chapter 5.05).

15.07 External works specification

The following specification and take-off for external works is designed with the intention of providing an example of the many different work sections that are required when booking dimensions for external works. The take-off and specification notes (below) should be read in conjunction with drawing ALXT200289.

15.07.01 Mandatory information

The works comprise the provision of a macadam access road with turning head, together with brick-paved hardstanding for four electric vehicle charging bays, an enclosed and gated bin store, a covered cycle secure storage enclosure together with soft landscaping including tree planting and turf. The whole comprises a rectangular site measuring 41 × 37 m on plan, and is enclosed on three sides by brick walling or close-boarded fencing. Please note that all work beyond the site boundary will be carried out by others.

15.07.02 Excavation and filling

The site is assumed to be level, with clear contractor access from the existing roadway. Ground conditions are stable, with no known hazardous materials present on site. Site survey information shows 150 mm topsoil overlaying sandstone, depths all as detailed on the structural engineer's borehole log dated 20/02/20. The groundwater level was established at 3.50 m below existing ground level on 20/02/20. No over or underground services cross the site.

Topsoil is to be excavated and stored on site in temporary spoil heaps. Bulk excavation to a depth of 500 mm to areas of brick-paved hardstanding and roadway and to a depth of 300 mm for areas of precast concrete paving. All bulk excavated materials to be removed from site.

15.07.03 Boundary walling and fencing

15.07.03.01 Boundary walling

Excavate foundation trench, minimum 600 mm deep. Weak mix concrete (C7) in blinding layer and in situ concrete (C20 20 mm agg) to strip foundation cross-sectional size 550 × 225mm.

A one-brick wall (215 mm thick) in Funtley red facing bricks laid English bond in cement mortar (1:3) pointed with bucket handle joints as work proceeds. Finished with brick-on-edge double bullnosed coping course in Southwater red engineering brick pointed in cement mortar (1:3) as work proceeds. Double bullnosed Southwater red engineering coping brick at corners size 215 × 215 × 103 mm pointed in cement mortar (1:3) as work proceeds.

Build-in 16 mm movement (expansion) joints at 11.500 m centres. Height above ground level: 1800 mm.

15.07.03.02 Fencing (part boundary surround and bin store enclosure)

Close-boarded fencing (1800 mm high) comprising 100 × 100 mm oak posts with weathered top @ 2700 mm centres set in 450 × 450 × 650 mm concrete (C15) on 50 mm thick course aggregate blinding, 75 × 75 mm oak triangular section arris rails morticed to fit posts, overlapping 90 × 14 mm tapered oak boards fixed vertically with galvanised nails to arris rails and 200 × 25 mm treated softwood gravel board.

Access to bin store: oak matchboard ledged and braced gate 1000 mm wide × 1800 mm high including 150 × 150 mm oak posts set in 450 × 450 × 650 mm concrete (C15) on 50 mm thick course aggregate blinding. Supply and fix the following ironmongery: one-and-a-half pairs of heavy-duty galvanised reversible gate hinges, heavy-duty galvanised Suffolk latch and heavy-duty galvanised barrel bolt.

15.07.04 Roadway

Wearing course of coated macadam (10 mm) on 60 mm thick base course of dense macadam laid to form falls and cross-falls on sub-base of 150 mm hardcore. Half-battered 125 × 255 mm precast concrete road kerb (BSEN1340) on and including 350 mm wide × 150 mm deep in situ concrete (C20) kerb foundation. Drop kerbs and channels to parking bay area (25,972 mm length).

15.07.05 Precast concrete paving

Precast concrete paving (900 × 600 × 63 mm; BS EN1339) bedded on cement and sand (1:4) to falls and cross-falls on 100 mm gravel base and 100 mm hardcore sub-base. 50 × 150 mm precast concrete rectangular section edging bedded on 150 mm wide × 100 mm deep in situ concrete (C20) paving edging foundation.

15.07.06 Brick paving (permeable pavement)

Permeable paving blocks herringbone pattern (215 × 103 × 80 mm) to BS EN1338:2003 laid in situ on 30 mm no-fines gravel to falls and cross-falls on 275 mm thick sub-base comprising 125 mm of 10–20 mm angular clean gravel on 150 mm free-draining hardcore. Brush finished with 2–4 mm grit into voids and between blocks.

15.07.07 Edging detail to brick paving

Precast concrete rectangular section edging (50 × 150 mm; BSEN 1340) bedded on 150 mm wide × 100 mm deep in situ concrete (C20) paving edging foundation.

15.07.08 Brick paving accessories

Recycled plastic moving wheel stop block 2500 mm long × 350 mm wide × 100 mm high fixed through sub-base with 600 mm long ground anchors.

15.07.09 Brick paving line marking

Road line marking to provide four number echelon (45°) parking bays each marked with following lettering 'EV CHARGE ONLY' (refer to drawing ALXT200289).

15.07.10 Equipment

Pedestal-mounted electrical vehicle charging points together with electrical cabling and associated electrical service equipment by others. Galvanised powder-coated cycle compound including base plate fixing by others.

15.07.11 Soft landscaping

Topsoil filling 150 mm deep obtained from temporary spoil heaps not exceeding 20 m from excavation as base to receive turf and as cultivated filling to planter.

Three number nursery stock trees (*Sorbus aucuparia*) BS size designation: heavy standard (HS).

Cultivate topsoil (75 mm deep) and surface apply a pre-turf mini-granular organic-based fertiliser and incorporate into the top 25 mm of the topsoil to receive turf size 610 × 1640 × 50 mm.

15.08 External works drawing

The take-off and specification notes should be read in conjunction with drawing ALXT200289.

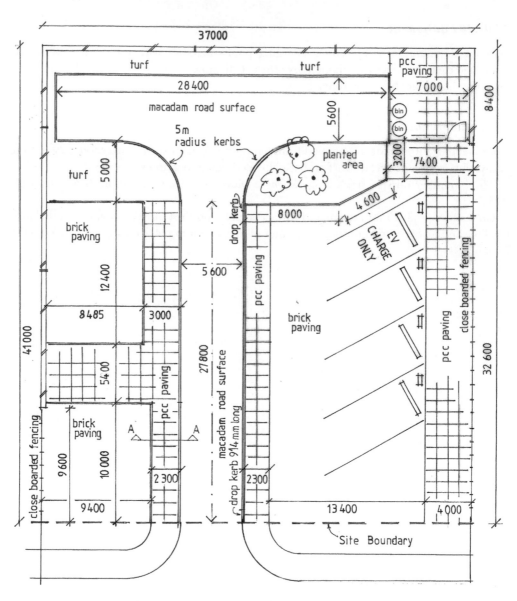

SITE PLAN   Drawing ref ALXT200289

To be read in conjunction with specification notes

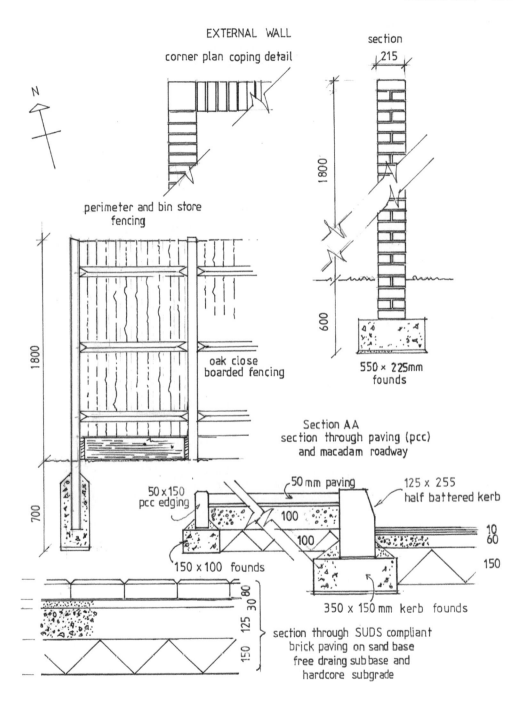

EXTERNAL WALL

corner plan coping detail

section

215

1800

600

550 × 225mm
founds

perimeter and bin store
fencing

N

1800

700

oak close
boarded fencing

Section AA
section through paving (pcc)
and macadam roadway

50 × 150
pcc edging

50 mm paving

125 × 255
half battered kerb

100

100

10
60

150

150 × 100 founds

350 × 150 mm kerb founds

80
30
125

150

section through SUDS compliant
brick paving on sand base
free draing sub base and
hardcore subgrade

15.09 External works sample take-off

The external works sample take-off is designed with the intention of demonstrating the many different work sections that are required when booking dimensions for external works.

External Works　–　Access road and parking　　Job ref: ALXT 200289　　　　Page 01 of 16

Mandatory Information:

The works comprise the provision of a macadam access road with turning head together with SUDS compliant brick paved hard standing for four electric vehicle charging bays, an enclosed and gated bin store, a covered cycle secure storage enclosure together with soft landscaping including tree planting and turf. The whole comprises a rectangular site measuring 41 metres x 37 metres on plan and is enclosed on three sides by brick walling or close boarded fencing. Note: all work beyond the site boundary, the cycle enclosure and pedestal mounted electrical vehicle charging points together with electrical cabling and associated electrical service equipment will be carried out by others.

Take off list

External Works

- Excavate topsoil

- Retain excavated materials in temporary spoil heaps on-site

- Bulk (reduce level) excavation to roadway
  - Disposal excavated materials off-site
  - Hardcore filling
  - Coated macadam roadway

- Bulk (reduce level) excavation to areas of brick paved hardstanding
  - Disposal excavated materials off-site
  - SUDS compliant hardcore filling
  - Coarse gravel imported filling
  - Interlocking brick paving

- Bulk (reduce level) excavation to areas of precast concrete paving
  - Disposal excavated materials off-site
  - Hardcore filling
  - Coarse gravel filling
  - Precast concrete slabs

- Boundary wall
  - Excavate foundation trench
  - Backfilling with excavated mats
  - Blinding concrete to bottoms of trench
  - Concrete in foundation trench
  - Adjust (ddt) backfilling and (add) disposal excavated mats
  - Brickwork in walls
  - Adjust (ddt) backfilling and (add) disposal excavated mats
  - EO for coping course
  - Special purpose corner coping bricks
  - Expansion (movement) joints

External Works – Access road and parking   Job ref: ALXT 200289   Page 02 of 16

Take off list -
(Contd)

- Fencing
  - Close boarded oak fencing
  - Extra over fencing for special supports
    - End posts
    - Corner posts
    - Integral gate posts
    - Gates and associated ironmongery
  - Foundation excavation for fence posts
  - Disposal excavated materials
  - Blinding bed
  - Concrete foundations to fence posts
- Kerbs,
  - straight
  - radius bend
  - EO last for drop kerbs
  - Adjust straight kerb for channel kerb
- Precast concrete path edging
- Turfing
  - Topsoil filling
  - Cultivation
  - Turf
  - Maintenance
  - Adjust topsoil away from site
- Planter
  - Topsoil filling
  - Cultivation
  - Adjust topsoil away from site
  - Trees
- Parking lines
- Parking bay lettering
- Wheel stops

Notes:-
1) Work beyond the boundary of the site will be costed and measured elsewhere.

2) Given the nature of the substrata the Structural Engineer has stipulated that temporary support to the sides and faces of open excavation will not be required.

3) It is assumed that the site has been cleared of all vegetation and clear contractor access is possible.

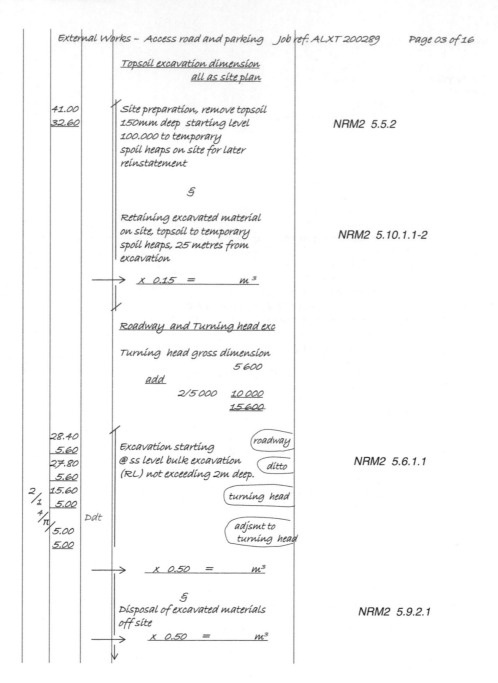

Topsoil excavation dimension
                all as site plan

41.00   Site preparation, remove topsoil   NRM2 5.5.2
32.60   150mm deep starting level
        100.000 to temporary
        spoil heaps on site for later
        reinstatement

                    &

        Retaining excavated material
        on site, topsoil to temporary   NRM2 5.10.1.1-2
        spoil heaps, 25 metres from
        excavation

            ⟶   x 0.15 =   m³

Roadway and Turning head exc

        Turning head gross dimension
                        5 600
            add
                2/5 000   10 000
                          15 600

28.40   Excavation starting   (roadway)
 5.60   @ ss level bulk excavation (ditto)   NRM2 5.6.1.1
27.80   (RL) not exceeding 2m deep.
 5.60   (turning head)
2/ 15.60
1/ 5.00   (adjsmt to
4/   Ddt   turning head)
π/ 5.00
 5.00

            ⟶   x 0.50 =   m³

                    &
        Disposal of excavated materials   NRM2 5.9.2.1
        off site
            ⟶   x 0.50 =   m³

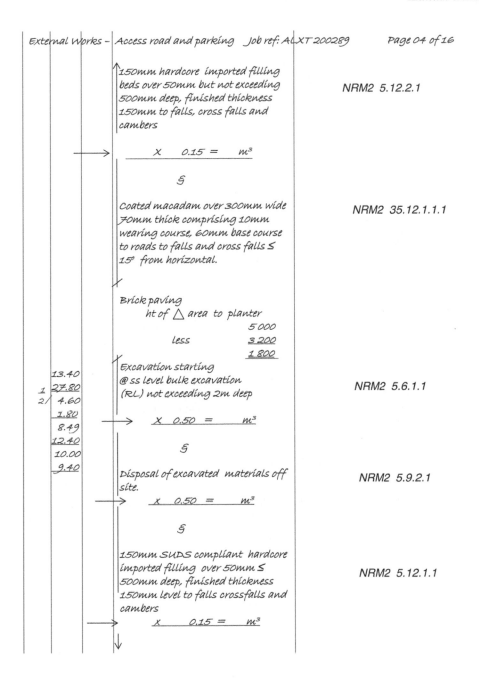

150mm hardcore imported filling
beds over 50mm but not exceeding   NRM2 5.12.2.1
500mm deep, finished thickness
150mm to falls, cross falls and
cambers

                    X   0.15 =   m³

                         &

Coated macadam over 300mm wide   NRM2 35.12.1.1.1
70mm thick comprising 10mm
wearing course, 60mm base course
to roads to falls and cross falls ≤
15° from horizontal.

Brick paving
          ht of △ area to planter
                                        5 000
                     less   3 200
                                        1 800

		Excavation starting
	13.40	@ ss level bulk excavation
1	27.80	(RL) not exceeding 2m deep
2/	4.60	
	1.80	
	8.49	
	12.40	
	10.00	
	9.40	

                                                  NRM2 5.6.1.1

                    X 0.50 =   m³

                         &

Disposal of excavated materials off   NRM2 5.9.2.1
site.
                    X   0.50 =   m³

                         &

150mm SUDS compliant hardcore
imported filling over 50mm ≤   NRM2 5.12.1.1
500mm deep, finished thickness
150mm level to falls crossfalls and
cambers
                    X   0.15 =   m³

External Works – Access road and parking   Job ref: ALXT 200289   Page 05 of 16

Brick paving (contd)

125mm coarse gravel imported
filling all abd BUT 125mm
finished thickness

NRM2 5.12.1.1

→   X   0.13 =   m³

§

Imported filling no fines gravel
as bedding, coarse < 50mm thick,
finished thickness 30mm level and
to falls, cross falls and cambers

NRM2 5.12.1.1

§

215 x 103 x 50mm thick
interlocking bricks (BSEN 1338;
2003) > 300mm wide in roads
and pavings, to falls cross falls
and slopes ≤ 15° from horizontal,
herringbone pattern, brush
finished with 2 -4mm grit.

NRM2 35.14.1.1

Notwithstanding NRM2 35.14.*.*.1
bedding has been measured
separately as filling.

Precast concrete paving

width of west area of pcc paving
                                    8 485
                                    3 000
                                   11 485
Length of 4m wide paving to east
boundary
                                   32 600
                                    3 200
                                   29 400

External Works – Access road and parking   Job ref: ALXT 200289   Page 06 of 16
Precast concrete paving (contd)

11.49	Excavation starting
5.40	@ ss level bulk excavation
12.40	(RL) not exceeding 2m deep

NRM2 5.6.1.1

X   0.30 =   m³

&

Disposal of excavated materials off site.

NRM2 5.9.2.1

3.00
10.00
2.30
27.80
2.30
29.40
4.00
8.40
7.00
7.40
3.20

X   0.30 =   m³

&

150mm SUDS compliant
hardcore imported filling all abd

NRM2 5.12.1.1

X   0.15 =   m³

&

Imported hardcore filling beds over
50mm thick ≤ 500mm deep,
finished thickness 100mm

NRM2 5.12.2.1

X   0.10 =   m³

&

Precast concrete paving slabs
900 x 600 x 63mm to BSEN1339
> 300mm to falls cross falls and
slopes ≤15° from horizontal.

NRM2 14.1.1.3

External Works – Access road and parking   Job ref: ALXT 200289   Page 07 of 16

<u>Boundary wall</u>

Excavation depth	600	
less topsoil	<u>150</u>	
	450	
	<u>Ext girth</u>	
	41 000	
less	<u>9 600</u>	
	31 400	
	37 000	
	<u>8 400</u>	
less 2 / 1	76 800	
2 / 215		
	<u>215</u>	
Centre line of ext wall	<u>76 585</u>	

<u>brickwork height</u>	1 800
	<u>600</u>
	2 400
less founds depth	<u>225</u>
height of bwk	<u>2 175</u>

76.59	Foundation excavation	
0.55	commencing at strip site level	NRM2 5.6.2.1
0.45	foundation excavation ≤ 2 m deep	

&

Filling obtained from excavated
materials, final thickness not
exceeding 500mm deep from
temporary spoil heaps 25m
from excavation

> Not withstanding NRM2 5.11.2 filling not exceeding 500mm deep has been measured in m³

	Plain in-situ weak mix concrete	
76.59	(C7) horizontal work ≤ 300 th in	NRM2 11.2.1.1.1
0.55	blinding poured on or against	
	earth or unblinded hardcore	

→   x   0.05 =   m³

&

Plain in-situ weak mix concrete
(C20) horizontal work in
≤ 300mm thick in structures

NRM2 11.2.1.2

x 0.23 = m^3

&

Deduct
Filling obtained from excavated
materials, final thickness not
exceeding 500mm deep from
temporary spoil heaps 25m
from excavation

NRM2 5.11.2

x 0.23 = m^3

&

Add
Disposal of excavated materials
off site

NRM2 5.9.2.1

x 0.23 = m^3

76.59	Walls 215mm th in Funtley red facing bricks laid English bond in cement mortar (1:3) pointed with bucket handle joints awp.
2.18	

NRM2 14.1.1

76.59   Extra over walls for perimeter work
215 x 103 x 65mm in Southwater
red double bullnose coping brick
pointed in c.m. (1:3) awp

NRM2 14.11.1.1

2   Special purpose made brick in
Southwater red double bullnosed
coping brick 215 x 215 x 103mm
pointed in c.m. (1:3) awp

NRM2 14.13.1.1

340   *External works*

			Adjustment for backfilling	
76.59			Ddt	
0.22			Backfilling	NRM2 5.11.2
0.23				
			&	
			Add	
			Disposal of excavated materials	
			off site	NRM2 5.9.2.1
			Vertical movement joints	
			@ 11.50 m centres	
			Centre Line of wall 76.585	
			joints @ centres of 11 500	
			= 6.69 which means 7 joints	
7	2.18		Expansion joints 16mm wide to	NRM2 14.22.1
			215mm thick solid wall	
			Fencing	
9.60			Close boarded oak fencing	NRM2 36.1.1.1
8.40			1800mm high comprising 100 x	
7.40			100mm oak posts with weathered	
32.60			top @ 2700mm centres set	
			including 3nr 75 x 75mm oak	
			triangular section arris rails	
			mortised to fit posts, overlapping	
			90 x 14mm tapered oak boards	
			fixed vertically with galvanised	
			nails to arris rails and 200 x 25	
			mm treated softwood gravel	
			board.	

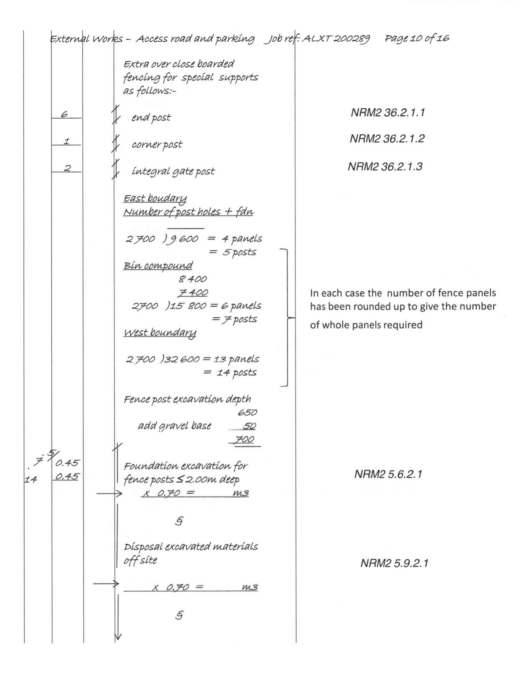

Extra over close boarded
fencing for special supports
as follows:-

6	end post	NRM2 36.2.1.1
1	corner post	NRM2 36.2.1.2
2	integral gate post	NRM2 36.2.1.3

<u>East boudary</u>
<u>Number of post holes + fdn</u>

2700) 9 600 = 4 panels
$\qquad$ = 5 posts

<u>Bin compound</u>
$\qquad$ 8 400
$\qquad$ 7 400
2700)15 800 = 6 panels
$\qquad$ = 7 posts

<u>West boundary</u>

2700)32 600 = 13 panels
$\qquad$ = 14 posts

In each case the number of fence panels
has been rounded up to give the number
of whole panels required

Fence post excavation depth
$\qquad$ 650
add gravel base $\qquad$ 50
$\qquad$ 700

    ⋅7 / ⁵⁄₁ 0.45
    14   0.45

Foundation excavation for
fence posts ≤ 2.00m deep   NRM2 5.6.2.1

→ x 0.70 =   m3

&

Disposal excavated materials
off site   NRM2 5.9.2.1

→ x 0.70 =   m3

&

342 *External works*

Blinding bed, course gravel ≤ 50mm thick, finished thickness 50mm

NRM2 5.12.1.1

x 0.05 =   m3

&

Mass concrete (C15) in filling around fence posts (26nr) poured on or against unblinded hardcore

NRM2 11.1.1.3.1

x 0.65 =   m3

1 | Gates 1000 wide x 1800mm high oak ledged and braced to match close boarded fencing.

NRM2 36.5.1.1

Supply and fix the following Ironmongery with galvanised screws to oak :

1 | 1½ pairs gate hinges pairs of heavy duty galvanised reversible 600mm long gate hinges

NRM2 36.6.1.1-2

1 | Suffolk gate latch and keep heavy duty galvanised Suffolk latch

NRM2 36.6.1.1-2

NRM2 36.6.1.1-2

1 | Heavy duty galvanised barrel bolt 225mm x 37mm with bolt keep.

<u>Kerbs (straight)</u>

	28 400
less 2/5 000 10 000	
5 600	15 600
	13 800

2 | 28.40
| 5.60
2 | 13.80
| 27.80

Precast concrete half battered kerb to BS 1340, size 125 x 255mm bedded and jointed in c.m. (1:3) haunched with in situ concrete (C10), on and including 350 wide x 150mm deep (C20) foundation

NRM2 35.1.1.*.1

Adjustment for channel @ parking bay

	27 800
less 2/914	1 828
	25 972

<u>Ddt</u>

25.97   Pcc straight kerbs abd

NRM2 35.1.1.*.1

§

<u>Add</u>

Precast concrete channel to BSEN 1340, size 914 x 305 x 150mm bedded and jointed in c.m. (1:3) and haunched with in situ concrete (C10) on and including 350 wide x 150mm deep (C20) foundation

NRM2 35.1.1.*.1

<u>Kerbs (radius bend)</u>

2/
1
4/2/π

5.00   Pcc channel abd but to curved bend on plan, radius 5000mm

NRM2 35.3.1.1.1

Extra over straight kerbs for special units;

1   left hand tapered drop kerb

NRM2 35.3.1.1

1   right hand tapered drop kerb

NRM2 35.3.1.1

344 *External works*

Pcc path edging
$$7\,400$$
less $\underline{4\,000}$
$$\underline{3\,400}$$

2/	32.60
	27.80
2/	12.40
2/	8.49
	9.40
	10.00
	13.40
	8.00
	4.60
	3.00

Precast concrete rectangular
section edging to BSEN 1340 , 50
x 150 mm bedded and jointed in
c.m. (1:3) , haunched with in-situ
concrete (C10) on and including
150 wide x 100mm deep (C20)
foundation.

(to planter)

NRM2 35.2.1.*.1

Turfing areas west of turning
head (width)
$$37\,000$$
less
$$28\,400$$
$$7\,000$$
2/215 $\underline{\quad 430\quad}$ $\underline{35\,830}$
$$\underline{1\,170}$$
Turfing areas west of turning
head (length)
$$5\,600$$
add   2/125 $\underline{\quad 250}$
$$\underline{5\,850}$$
Turf to north of turning head

Length   37 000
less $\underline{\quad 8\,400}$
$$\underline{29\,600}$$

Width   8 400
less $\underline{\quad 5\,600}$
$$\underline{2\,800}$$

External Works – Access road and parking   Job ref: ALXT 200289   Page 14 of 16

<u>Turf to curved area west of</u>
<u>turning head</u>

	8 485
	3 000
	11 485
less	5 000
	6 485

	29.60	Final thickness of filling not exceeding 500mm deep, finished thickness 150mm, topsoil from temporary spoil heaps 20 metres from filling to form turf beds	NRM2 11.1.1
	2.80		
	5.85		
	1.17		
1	6.49	&	
⁴/π	5.00		
	5.00		
	5.00	Cultivation 75 mm deep to achieve fine tilth and surface apply a pre-turf fertilizer mini granular organic based fertilizer and incorporate into the top 25mm of the topsoil.	NRM2 37.4.1

&

General amenity turf size 610mm x 1640mm x 50mm thick   NRM2 37.1.1

<u>Topsoil filling to planter</u>

<u>Length</u> of planter	4 000
	13 400
	2 300
	19 700
less	7 400
	12 300

<u>width</u> of planter	5 000
less straight width	3 200
height of triang cut out	1 800

	12 300
Length of quarter circle	5 000
remaining length	7 300

346 *External works*

1	7.30		Final thickness of filling not exceeding 500mm deep, finished thickness 150mm, topsoil from temporary spoil heaps 20 metres from filling to form planter beds	NRM2 11.1.1
4/π	5.00			
	5.00			
	5.00			
1		Ddt		
2/	1.80			
	4.30			

&

Cultivation 150mm deep manual cultivation to achieve fine tilth.   NRM2 37.1.1

3   Nursery stock trees (Sorbus aucuparia) BS size designation: heavy standard (HS).   NRM2 37.5.1.1

5/ 6.00   Line marking width ≤300mm, straight, hot applied thermoplastic, white to brick paved surface   NRM2 35.23.1.1.2

4   Letters, figures and symbols hot applied thermoplastic, white, 'EV CHARGE ONLY' to brick paved surface.   NRM2 35.24.1

4	Extra over brick paving for accessories , recycled plastic moving wheel stop block 2500 long x 350 wide x 100mm high fixed through brick paving and sub base with 600mm long ground anchors.	NRM2 35.15.2.1
	Note : the following equipment and accessories to be provided by others..:-	
	Pedestal mounted electrical vehicle charging points together with electrical cabling and associated electrical service equipment.	
	Galvanised powder coated cycle compound approximately 12m x 8m including base plate to brick paving surface.	

Appendix 1

Common abbreviations used when booking dimensions

a.b.	as before	cc.	curved cutting
a.b.d.	as before described	C.I.	cast iron
agg.	aggregate	clg.jst.	ceiling joist
a.f.	after fixing	c.jtd.	close jointed
asph.	asphalt	C.P.	chromium plated
av.	average	c.o.e.	curved on elevation
av.g.l.	average ground level	c.o.p.	curved on plan
awp	as work proceeds	c.s.g.	clear sheet glass
b. & p.	bed & point	c.t. & b.	cut tooth and bond
b.e.	both edges	chfd.	chamfered
b.f.	before fixing	chy.	chimney
b.i.	build in	C.L.	centre line — · — · — · —
b.m.	birdsmouth	clg.	ceiling
b.n.	bull nosed	col.	column
b.s.	both sides	cos.	course
bal.	baluster	cpd.	cupboard
basmt.	basement	conc.	concrete
bdd.	bedded	csk.	countersunk
bdg.	boarding	cmt.	cement
bk.	brick	d/d	delivered
bkt.	bracket	Ddt.	deduct
bldg.	building	d.h.	double hung
brd.	board	dia. *or* ϕ	diameter
brrs.	bearers	dist.	distance
bwk. *or* bkk.	brickwork	D.P.C.	damp proof course
b.o.e.	brick on edge/end	D.P.M.	damp proof membrane
B.S.	British Standard	d.p.	distance piece
B.S.E.N.	British Standard Eurocodes	dp.	deep
casmt.	casement	E.G.L.	existing ground level
cav.	cavity	E.M.L.	expanded metal lathing
cav .ins.	cavity insulation	E.O.	extra over
c. & f.	cut and fit	E.S.	earthwork support
c. & p.	cut and pin	ea.	each
c. & s.	cement and sand	exc.	excavate
c.b.	common bricks	exen.	excavation
c.bwk	common brickwork	extl.	external
cc.	centres	extg.	existing

F.A.I.	fresh air inlet	m^2	square metre
f.c.	fair cutting	m^3	cubic metre
f.f.	fair face	matl.	material
f. & b.	framed and braced	m.g.	make good
f.l. & b.	framed ledged and braced	M.H.	manhole
F.L.	floor level	M.S.	mild steel
fcgs.	facings	m.s.	measured separately
fdns.	foundations	mm	millimetre
fin.	finished	mit.	mitres
fr.	frame	mo.	moulded
frd.	framed	mort.	mortice
fwk.	formwork	msd.	measured
ftd.	fitted		
		n.e. or <	not exceeding
G.F.	ground floor	Nr.	number
G.I.	galvanised iron		
G.L.	ground level	o/a	overall
galv.	galvanised	o.c.n.	open copper nailing
grano.	granolithic	o.s.	one side
gth.	girth	opg.	opening
		orgl.	original
h.b.s.	herring bone strutting		
h.b.w. *or* ½ b.w.	half brick wall		
hdb.	hardboard	pbd.	plasterboard
h.c.	hardcore	P.C. sum	Prime Cost sum
hdg.jt.	heading joint	p. & s.	plank and strut
h.m.	hand made	plas.	plaster
hoz.	horizontal	plasd.	pastered
H.P.	high pressure	p.m.	purpose made
h.r.	half round	p.o.	prime only
h.t.	hollow tile	pol.	polished
ht.	height	pr.	pair
hw.	hardwood	Prov. sum	Provisional sum
		prep.	prepare
inc.	including	pt.	point
ins.	insulation	ptd.	pointed
intl.	internal	ptg.	pointing
inv.	invert	ptn.	partition
I.C.	inspection chamber	PVCu	unplasticized polyvinyl chloride
Jap.	Japanned	pvg.	paving
jst.	joist		
jt.	joint	r. & s.	render and set
jtd.	jointed	r.f. & s.	render float and set
		rad.	radius
K.P.S.	knot, prime, stop	R.C.	reinforced concrete
		r.c.	raking cutting
Lab.	labour	rdd.	rounded
l. & b.	ledged & braced	reinf.	reinforced or reinforcement
l.p.	large pipe	R.E.	rodding eye
l. & c.	level and compact	R.L.	reduced levels
		r.l.jt.	red lead joint
m	metre	r.m.e.	returned mitred end
		r.o.j.	rake out joint

R.S.C.	rolled steel channel
R.S.J.	rolled steel joist
R.W.H.	rainwater head
R.W.P.	rainwater pipe
reb.	rebated
retd.	returned
ro.	rough
S.A.A.	satin anodised aluminium
s.b.j.	soldered branch joint
s.d.	screw down
s.c.	stop cock
segtl.	segmental
s.e.	stopped end
s.g.	salt glazed
s.jt.	soldered joint
s.l.	short length
soff.	soffit
s.p.	small pipe
s.q.	small quantities
s.w.	softwood
sk.	sunk
sktg.	skirting
sq.	square
s. & l.	spread and level
S. & V.P.	soil and vent pipe
stg.	starting
swd.	softwood
T.	tee
T. & G.	tongued and grooved
t.	tonne
t. & r.	treads and risers
t.c.	terra cotta
t.p.	turning piece
tops.	topsoil
U.B.	universal beam
uPVC	unplasticized polyvinyl chloride

V.O.	variation order
V.P.	vent pipe
wi. or $\overline{w}$	with
w.g.	white glazed
W.I.	wrought iron
W.P.	waste pipe
wdw.	window
wthd.	weathered
X grain	cross grain
X tdg.	cross tongued

Symbols and abbreviations used in NRM2

ha	hectare
hr	hour
kg	kilogram
kN	kilonewton
kW	kilowatt
m	linear metre
m2	square metre
m3	cubic metre
mm	millimetre
mm2	square millimetre
mm3	cubic millimetre
nr	number
t	tonne
wk	week

Mathematical symbols

>	exceeding
≥	equal to or exceeding
≤	not exceeding
<	less than
%	percentage

Appendix 2

Geometric formulae

Title	Figure	Area	Perimeter
Rectangle		lb	$2(l+b)$
Parallelogram		lh	$2(l+b)$
Trapezium		$0.5h(a+b)$	$a+b+c+d$
Triangle		$0.5bh$ $0.5ab \sin C$ $0.5ac \sin B$ $0.5bc \sin A$ $\sqrt{[s(s-a)(s-b)(s-c)]}$ $[s = 0.5(a+b+c)]$	$a+b+c$
Circle		πr^2 $\dfrac{\pi D^2}{4}$	$2\pi r$ πD
Sector		$0.5r^2\theta$ [θ in radians] $0.5rl$ $\dfrac{\pi r^2 \theta}{360}$ [θ in radians]	arc length $l = r\theta$ [θ in radians]
Segment		$0.5r^2(\theta - \sin\theta)$ [θ in radians]	arc length $l = r\theta$ [θ in radians]
Ellipse		πab	$\pi(a+b)$

Title	Figure	Surface area	Volume
Cuboid		$2(ab+bc+ac)$	abc
Pyramid		$(a+b)l+ab$	$\frac{1}{3}abh$
Frustrum of a pyramid		$l(a+b+c+d) + \sqrt{(ab+cd)}$ [regular figure only]	$\dfrac{h}{3}(ab+cd+\sqrt{abcd})$
Wedge		Area $ABC'D'$ = area $ABCD \times$ $\cos\theta$ area $ABCD = \dfrac{\text{Area } ABC'D'}{\cos\theta}$	area $BCC' \times DC$
Cylinder		$2\pi rh+2\pi r^2$	πr^2h
Cone		$\pi rl+\pi r^2$ [total surface area]	$\frac{1}{3}\pi r^2h$
Frustrum of cone		$\pi r^2+\pi R^2+\pi(R+r)$ [total surface area]	$\dfrac{\pi h}{3}(R^2+Rr+r^2)$
Sphere		$4\pi r^2$	$\dfrac{4}{3}\pi r^2$
Segment of sphere		$2\pi Rh$	$\dfrac{\pi h}{6}(3r^2+h^2)$ $\dfrac{\pi h^2}{3}(3R-h)$

Index